ATLAS OF THE
ARAB-ISRAELI CONFLICT

Sixth edition

Martin Gilbert

Fellow of Merton College, Oxford

New York

OXFORD UNIVERSITY PRESS

1993

© 1974, 1976, 1979, 1984, 1992, 1993 Martin Gilbert

First published in Great Britain by
The Orion Publishing Group Limited
5 Upper St. Martin's Lane, London WC2H 9EA

Published in the United States of America by
Oxford University Press, Inc.
200 Madison Avenue
New York, N.Y. 10016, U.S.A.

Oxford is a registered trademark of
Oxford University Press

Library of Congress
Cataloging-in-Publication Data
Gilbert, Martin, 1936–
      [Arab-Israeli conflict]
      Atlas of the Arab-Israeli conflict / Martin Gilbert
         p.   cm.
      Includes bibliographical references and index.
      ISBN 0–19–521042–5 (hardback)
      ISBN 0–19–521062–X (paperback)
      1. Israel-Arab conflicts—Maps   2. Jewish-Arab relations—Maps.
      3. Palestine—Historical geography—Maps.   I. Title.
G2236.SIG52   1993   <G&M>
911'. 56—dc20
                                                    93–21921
                                                        CIP
                                                        MAP

Printing (last digit): 9 8 7 6

Printed in Great Britain

Preface

In this atlas I have traced the history of the Arab–Jewish conflict from the turn of the century to the present day. I have tried to show something of the intensity and bitterness of the conflict, of the types of incidents which it provoked, and of the views of those involved in it.

The majority of the maps in this atlas depict wars, conflict and violence, which have brought terrible suffering to all those caught up in them – Jew and Arab, soldier and civilian, adult and child. But there are also maps which show the various attempts to bring the conflict to an end, through proposals for agreed boundaries, through the signing of cease-fire agreements, and through negotiations. No map can show how peace will come, but they do show how much it is needed.

Since the publication of the last edition, my cartographer Terry Bicknell has died. He was a good friend as well as a fine draughtsman. The new maps in this edition have been drawn by Tim Aspden, Geoffrey Sims and Laura Sylvester, to whom my thanks are due. I am also grateful to Enid Wurtman, Abe Eisenstat and Burt Keimach for their help on many matters of detail. I should of course be grateful for readers' suggestions for further maps, for extra material, and for corrections.

Nineteen years have now passed since the first edition of this atlas. The subsequent new editions, including this one, have needed fifty-two maps to bring the story from the immediate aftermath of the war of October 1973 to the Arab uprising (Intifada) that began in 1987, the upsurge in Jewish immigration from Russia in 1989, the Gulf War of 1991, and the peace process inaugurated at Madrid in 1991, and revived following the visits of two successive United States Secretaries of State, James Baker (1991 and 1992) and Warren Christopher (1993).

The origins and earlier course of the conflict are often overshadowed in the public mind by the events of the past decade. But it is still my hope that, seen as a whole, this atlas will help to clarify the conflict throughout its lengthening span.

25 March 1993

MARTIN GILBERT
Merton College, Oxford

Maps

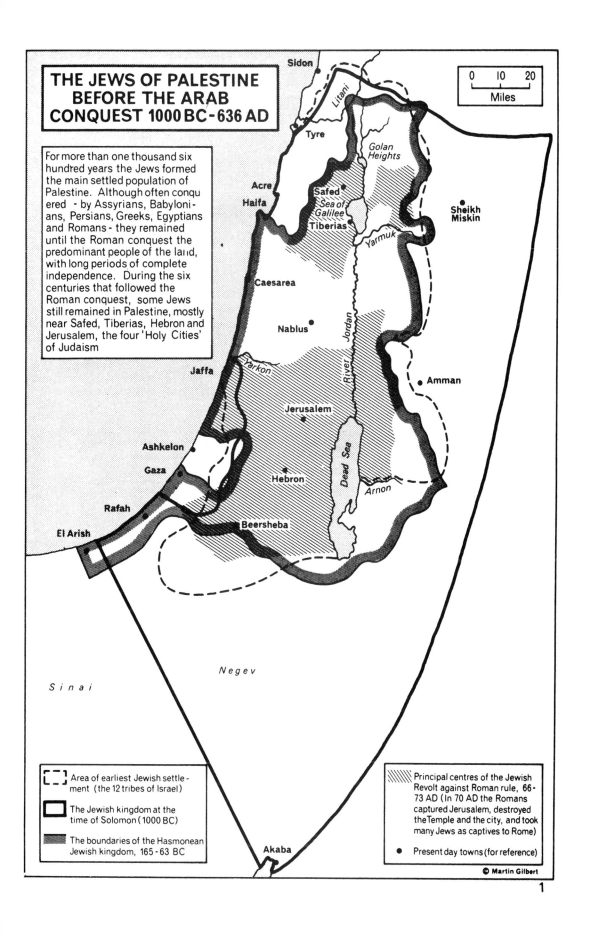

THE JEWS OF PALESTINE BEFORE THE ARAB CONQUEST 1000 BC - 636 AD

For more than one thousand six hundred years the Jews formed the main settled population of Palestine. Although often conquered - by Assyrians, Babylonians, Persians, Greeks, Egyptians and Romans - they remained until the Roman conquest the predominant people of the land, with long periods of complete independence. During the six centuries that followed the Roman conquest, some Jews still remained in Palestine, mostly near Safed, Tiberias, Hebron and Jerusalem, the four 'Holy Cities' of Judaism

Scale: 0   10   20 Miles

Sidon

Tyre

Golan Heights

Acre

Safed

Haifa

Sea of Galilee

Sheikh Miskin

Tiberias

Yarmuk

Caesarea

Jordan River

Nablus

Amman

Yarkon

Jaffa

Jerusalem

Ashkelon

Dead Sea

Gaza

Hebron

Arnon

Rafah

Beersheba

El Arish

Negev

Sinai

Akaba

- - - Area of earliest Jewish settlement (the 12 tribes of Israel)

▢ The Jewish kingdom at the time of Solomon (1000 BC)

▬ The boundaries of the Hasmonean Jewish kingdom, 165 - 63 BC

▨ Principal centres of the Jewish Revolt against Roman rule, 66-73 AD (In 70 AD the Romans captured Jerusalem, destroyed the Temple and the city, and took many Jews as captives to Rome)

● Present day towns (for reference)

© Martin Gilbert

1

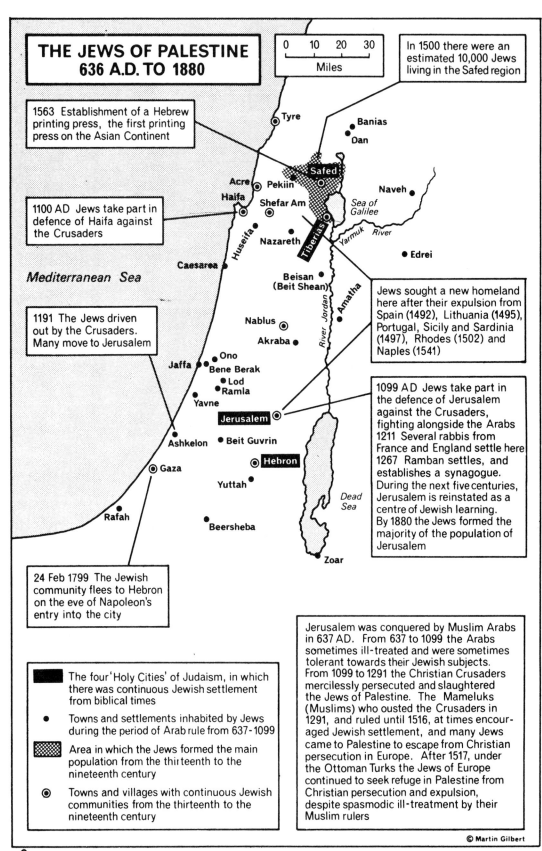

THE JEWS OF PALESTINE 636 A.D. TO 1880

0 10 20 30
Miles

In 1500 there were an estimated 10,000 Jews living in the Safed region

1563 Establishment of a Hebrew printing press, the first printing press on the Asian Continent

1100 AD Jews take part in defence of Haifa against the Crusaders

Mediterranean Sea

1191 The Jews driven out by the Crusaders. Many move to Jerusalem

Tyre

Banias
Dan

Safed

Naveh

Acre Pekiin
Haifa Shefar Am

Sea of Galilee

Huseifa Nazareth Tiberias

Yarmuk River

Caesarea

Edrei

Beisan (Beit Shean)

Amatha

Nablus

Akraba

River Jordan

Jews sought a new homeland here after their expulsion from Spain (1492), Lithuania (1495), Portugal, Sicily and Sardinia (1497), Rhodes (1502) and Naples (1541)

Ono
Jaffa Bene Berak
Lod
Ramla
Yavne

Jerusalem

Ashkelon Beit Guvrin

Hebron

Gaza

Yuttah

Rafah

Beersheba

Dead Sea

Zoar

1099 A D Jews take part in the defence of Jerusalem against the Crusaders, fighting alongside the Arabs 1211 Several rabbis from France and England settle here 1267 Ramban settles, and establishes a synagogue. During the next five centuries, Jerusalem is reinstated as a centre of Jewish learning. By 1880 the Jews formed the majority of the population of Jerusalem

24 Feb 1799 The Jewish community flees to Hebron on the eve of Napoleon's entry into the city

The four 'Holy Cities' of Judaism, in which there was continuous Jewish settlement from biblical times

● Towns and settlements inhabited by Jews during the period of Arab rule from 637-1099

Area in which the Jews formed the main population from the thirteenth to the nineteenth century

◉ Towns and villages with continuous Jewish communities from the thirteenth to the nineteenth century

Jerusalem was conquered by Muslim Arabs in 637 AD. From 637 to 1099 the Arabs sometimes ill-treated and were sometimes tolerant towards their Jewish subjects. From 1099 to 1291 the Christian Crusaders mercilessly persecuted and slaughtered the Jews of Palestine. The Mameluks (Muslims) who ousted the Crusaders in 1291, and ruled until 1516, at times encouraged Jewish settlement, and many Jews came to Palestine to escape from Christian persecution in Europe. After 1517, under the Ottoman Turks the Jews of Europe continued to seek refuge in Palestine from Christian persecution and expulsion, despite spasmodic ill-treatment by their Muslim rulers

© Martin Gilbert

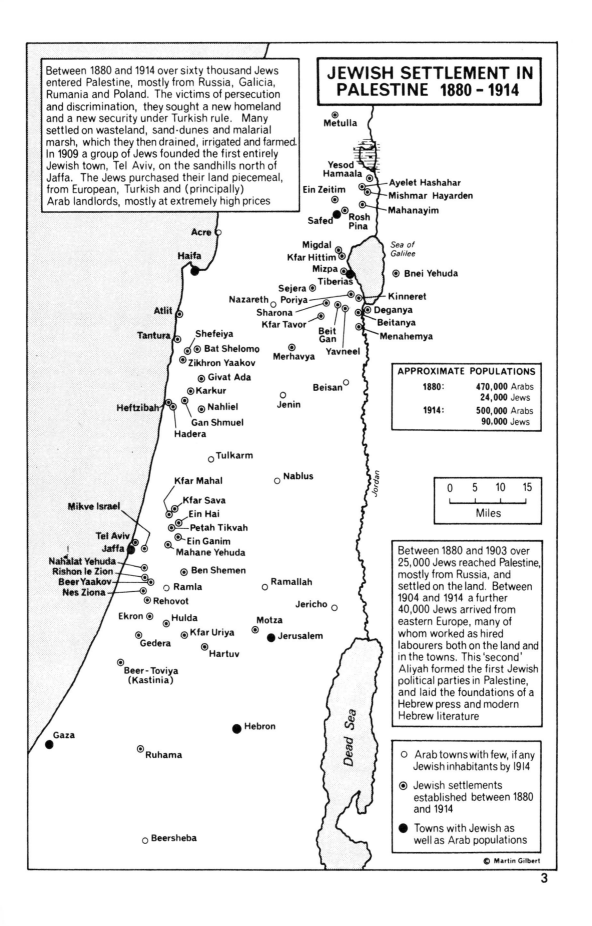

JEWISH SETTLEMENT IN PALESTINE 1880 – 1914

Between 1880 and 1914 over sixty thousand Jews entered Palestine, mostly from Russia, Galicia, Rumania and Poland. The victims of persecution and discrimination, they sought a new homeland and a new security under Turkish rule. Many settled on wasteland, sand-dunes and malarial marsh, which they then drained, irrigated and farmed. In 1909 a group of Jews founded the first entirely Jewish town, Tel Aviv, on the sandhills north of Jaffa. The Jews purchased their land piecemeal, from European, Turkish and (principally) Arab landlords, mostly at extremely high prices

Metulla

Yesod Hamaala

Ayelet Hashahar

Ein Zeitim

Mishmar Hayarden

Mahanayim

Safed

Rosh Pina

Acre

Migdal

Sea of Galilee

Haifa

Kfar Hittim

Mizpa

Tiberias

Bnei Yehuda

Sejera

Kinneret

Nazareth

Poriya

Atlit

Sharona

Deganya

Kfar Tavor

Beitanya

Tantura

Shefeiya

Beit Gan

Menahemya

Bat Shelomo

Merhavya

Yavneel

Zikhron Yaakov

Givat Ada

Beisan

Karkur

Jenin

Heftzibah

Nahliel

Gan Shmuel

Hadera

Tulkarm

Kfar Mahal

Nablus

Mikve Israel

Kfar Sava

Ein Hai

Petah Tikvah

Tel Aviv

Ein Ganim

Jaffa

Mahane Yehuda

Nahalat Yehuda

Ben Shemen

Rishon le Zion

Ramallah

Beer Yaakov

Ramla

Nes Ziona

Jericho

Rehovot

Ekron

Hulda

Motza

Kfar Uriya

Jerusalem

Gedera

Hartuv

Beer - Toviya (Kastinia)

Hebron

Gaza

Ruhama

Dead Sea

Beersheba

Jordan

APPROXIMATE POPULATIONS

1880:	470,000 Arabs
	24,000 Jews
1914:	500,000 Arabs
	90,000 Jews

0	5	10	15

Miles

Between 1880 and 1903 over 25,000 Jews reached Palestine, mostly from Russia, and settled on the land. Between 1904 and 1914 a further 40,000 Jews arrived from eastern Europe, many of whom worked as hired labourers both on the land and in the towns. This 'second' Aliyah formed the first Jewish political parties in Palestine, and laid the foundations of a Hebrew press and modern Hebrew literature

○ Arab towns with few, if any Jewish inhabitants by 1914

◉ Jewish settlements established between 1880 and 1914

● Towns with Jewish as well as Arab populations

© Martin Gilbert

3

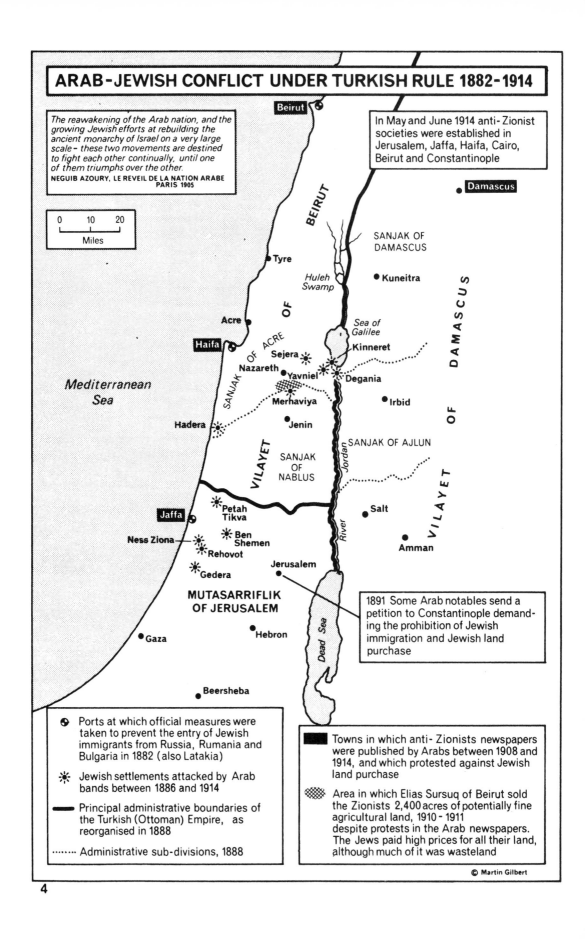

ARAB-JEWISH CONFLICT UNDER TURKISH RULE 1882-1914

The reawakening of the Arab nation, and the growing Jewish efforts at rebuilding the ancient monarchy of Israel on a very large scale – these two movements are destined to fight each other continually, until one of them triumphs over the other.

NEGUIB AZOURY, LE REVEIL DE LA NATION ARABE PARIS 1905

In May and June 1914 anti-Zionist societies were established in Jerusalem, Jaffa, Haifa, Cairo, Beirut and Constantinople

Beirut

Damascus

0 10 20
Miles

VILAYET OF BEIRUT

SANJAK OF DAMASCUS

Tyre

Huleh Swamp

Kuneitra

Acre

SANJAK OF ACRE

Sea of Galilee

Haifa

Sejera

Kinneret

Nazareth

Yavniel

Degania

Mediterranean Sea

Merhaviya

Irbid

Hadera

Jenin

VILAYET OF DAMASCUS

SANJAK OF AJLUN

SANJAK OF NABLUS

Jordan River

Salt

Petah Tikva

Ben Shemen

Ness Ziona

VILAYET

Rehovot

Amman

Gedera

Jerusalem

MUTASARRIFLIK OF JERUSALEM

1891 Some Arab notables send a petition to Constantinople demanding the prohibition of Jewish immigration and Jewish land purchase

Dead Sea

Gaza

Hebron

Beersheba

⊕ Ports at which official measures were taken to prevent the entry of Jewish immigrants from Russia, Rumania and Bulgaria in 1882 (also Latakia)

☀ Jewish settlements attacked by Arab bands between 1886 and 1914

━ Principal administrative boundaries of the Turkish (Ottoman) Empire, as reorganised in 1888

······· Administrative sub-divisions, 1888

▮ Towns in which anti-Zionist newspapers were published by Arabs between 1908 and 1914, and which protested against Jewish land purchase

▨ Area in which Elias Sursuq of Beirut sold the Zionists 2,400 acres of potentially fine agricultural land, 1910-1911 despite protests in the Arab newspapers. The Jews paid high prices for all their land, although much of it was wasteland

© Martin Gilbert

4

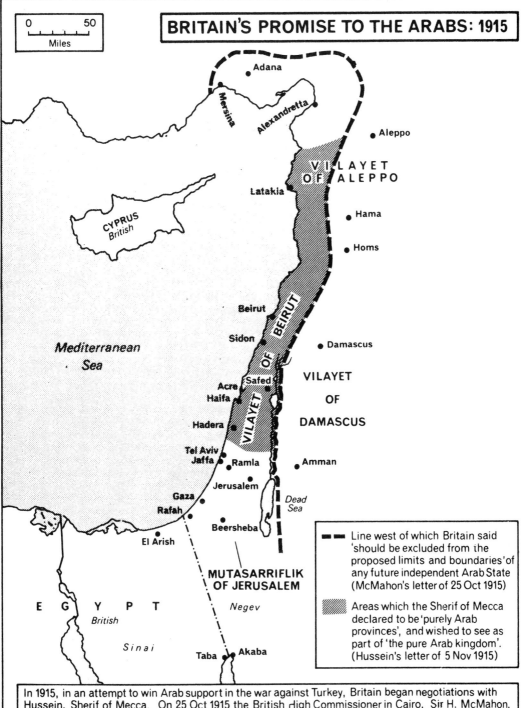

BRITAIN'S PROMISE TO THE ARABS: 1915

0 ___ 50
Miles

Adana

Mersina

Alexandretta

Aleppo

VILAYET OF ALEPPO

Latakia

Hama

Homs

CYPRUS
British

Beirut

VILAYET OF BEIRUT

Sidon

Damascus

Mediterranean Sea

Safed

VILAYET OF DAMASCUS

Acre
Haifa

Hadera

Tel Aviv
Jaffa

Ramla

Amman

Jerusalem

Dead Sea

Gaza
Rafah

Beersheba

El Arish

MUTASARRIFLIK
OF JERUSALEM

E G Y P T
British

Negev

Sinai

Taba   Akaba

--- Line west of which Britain said 'should be excluded from the proposed limits and boundaries' of any future independent Arab State (McMahon's letter of 25 Oct 1915)

////// Areas which the Sherif of Mecca declared to be 'purely Arab provinces', and wished to see as part of 'the pure Arab kingdom'. (Hussein's letter of 5 Nov 1915)

In 1915, in an attempt to win Arab support in the war against Turkey, Britain began negotiations with Hussein, Sherif of Mecca. On 25 Oct 1915 the British High Commissioner in Cairo, Sir H. McMahon, informed Hussein that Britain was 'prepared to recognize and support the independence of the Arabs....' But, he added, the Eastern Mediterranean littoral would have to be entirely excluded from any future Arab State. In his reply on 5 Nov 1915, Hussein insisted on the inclusion of the Vilayet of Beirut, but made no mention of the Mutasarriflik of Jerusalem. But on 14 Dec 1915 McMahon replied that any such inclusion 'will require careful consideration'. On 1 Jan 1916 Hussein warned McMahon: 'the people of Beirut will decidedly never accept such isolations'. At no point in the correspondence was any mention made of southern Palestine, Jerusalem or the Jews

© Martin Gilbert

5

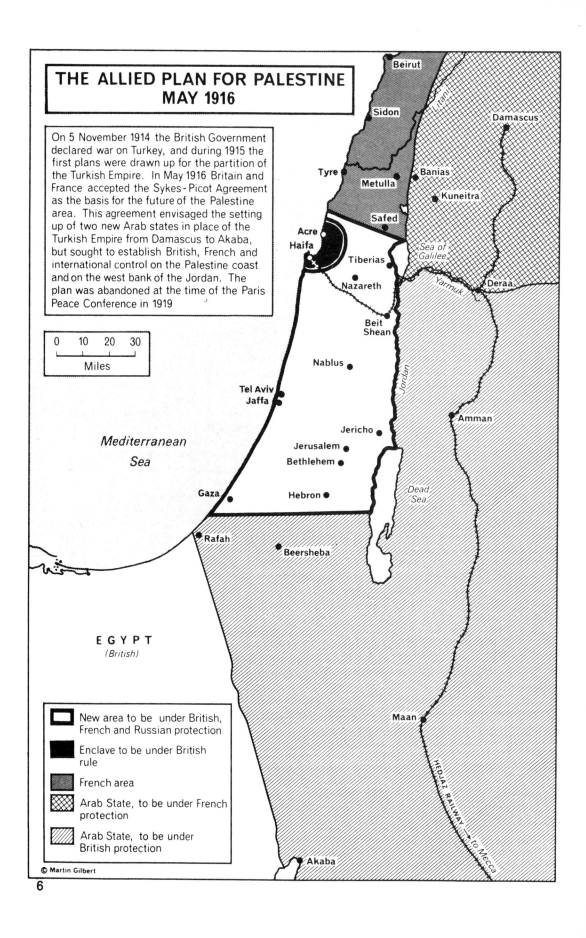

THE ALLIED PLAN FOR PALESTINE MAY 1916

On 5 November 1914 the British Government declared war on Turkey, and during 1915 the first plans were drawn up for the partition of the Turkish Empire. In May 1916 Britain and France accepted the Sykes-Picot Agreement as the basis for the future of the Palestine area. This agreement envisaged the setting up of two new Arab states in place of the Turkish Empire from Damascus to Akaba, but sought to establish British, French and international control on the Palestine coast and on the west bank of the Jordan. The plan was abandoned at the time of the Paris Peace Conference in 1919

0   10   20   30

Miles

Beirut

Sidon

Damascus

Litani

Tyre

Banias

Metulla

Kuneitra

Safed

Acre

Haifa

Sea of Galilee

Tiberias

Yarmuk

Deraa

Nazareth

Beit Shean

Jordan

Nablus

Tel Aviv

Jaffa

Amman

Mediterranean Sea

Jericho

Jerusalem

Bethlehem

Dead Sea

Gaza

Hebron

Rafah

Beersheba

E G Y P T
(British)

Maan

New area to be under British, French and Russian protection

Enclave to be under British rule

French area

Arab State, to be under French protection

Arab State, to be under British protection

HEDJAZ RAILWAY → to Mecca

© Martin Gilbert

Akaba

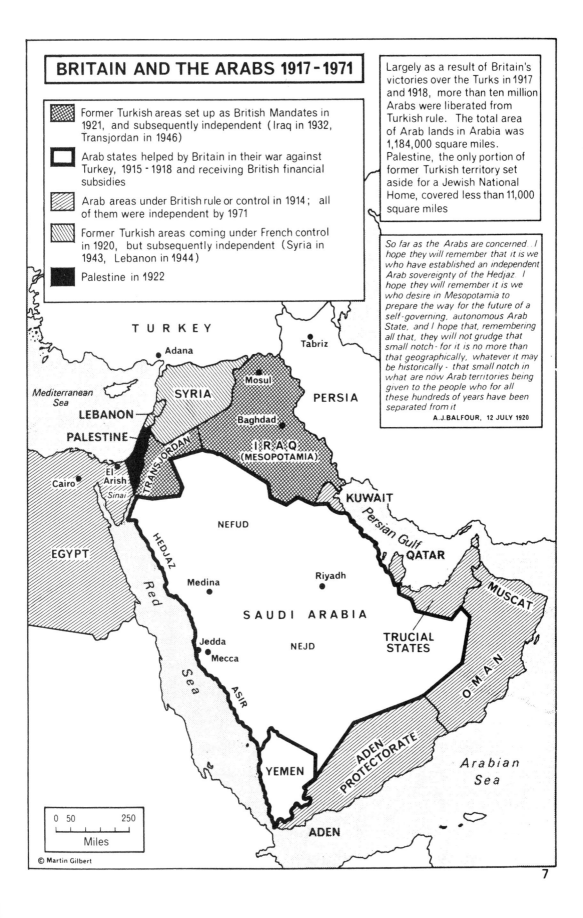

BRITAIN AND THE ARABS 1917-1971

Former Turkish areas set up as British Mandates in 1921, and subsequently independent (Iraq in 1932, Transjordan in 1946)

Arab states helped by Britain in their war against Turkey, 1915 - 1918 and receiving British financial subsidies

Arab areas under British rule or control in 1914; all of them were independent by 1971

Former Turkish areas coming under French control in 1920, but subsequently independent (Syria in 1943, Lebanon in 1944)

Palestine in 1922

Largely as a result of Britain's victories over the Turks in 1917 and 1918, more than ten million Arabs were liberated from Turkish rule. The total area of Arab lands in Arabia was 1,184,000 square miles. Palestine, the only portion of former Turkish territory set aside for a Jewish National Home, covered less than 11,000 square miles

So far as the Arabs are concerned...I hope they will remember that it is we who have established an independent Arab sovereignty of the Hedjaz. I hope they will remember it is we who desire in Mesopotamia to prepare the way for the future of a self-governing, autonomous Arab State, and I hope that, remembering all that, they will not grudge that small notch - for it is no more than that geographically, whatever it may be historically - that small notch in what are now Arab territories being given to the people who for all these hundreds of years have been separated from it

A.J.BALFOUR, 12 JULY 1920

TURKEY

Adana

Tabriz

Mosul

Mediterranean Sea

SYRIA

PERSIA

LEBANON

Baghdad

PALESTINE

TRANSJORDAN

I·R·A·Q (MESOPOTAMIA)

El Arish

Cairo

Sinai

KUWAIT

Persian Gulf

EGYPT

NEFUD

QATAR

HEDJAZ

Red

Medina

Riyadh

MUSCAT

Sea

SAUDI ARABIA

TRUCIAL STATES

Jedda

Mecca

NEJD

O·M·A·N

ASIR

Arabian Sea

ADEN PROTECTORATE

YEMEN

0 50   250

Miles

ADEN

© Martin Gilbert

7

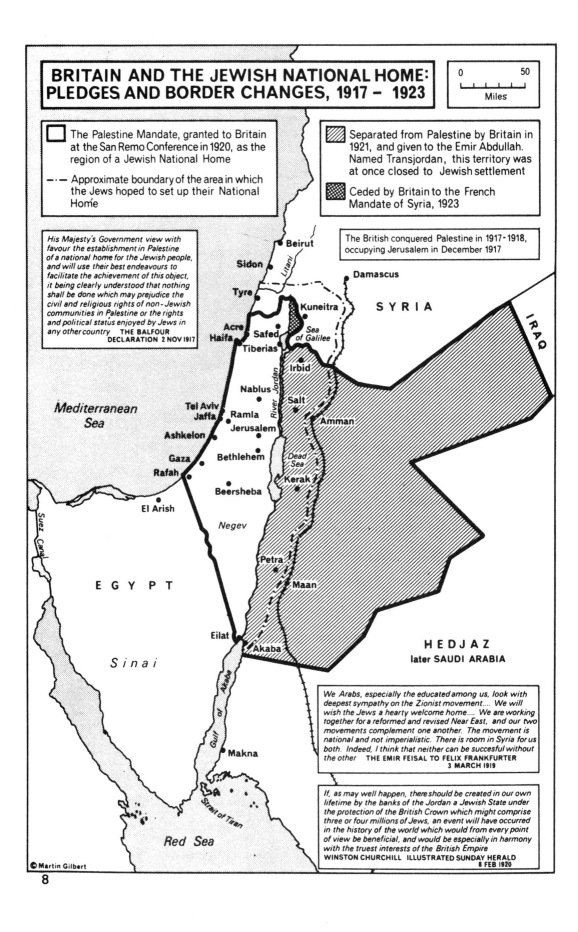

BRITAIN AND THE JEWISH NATIONAL HOME: PLEDGES AND BORDER CHANGES, 1917 – 1923

0   50
Miles

☐ The Palestine Mandate, granted to Britain at the San Remo Conference in 1920, as the region of a Jewish National Home

–·– Approximate boundary of the area in which the Jews hoped to set up their National Home

▨ Separated from Palestine by Britain in 1921, and given to the Emir Abdullah. Named Transjordan, this territory was at once closed to Jewish settlement

▨ Ceded by Britain to the French Mandate of Syria, 1923

His Majesty's Government view with favour the establishment in Palestine of a national home for the Jewish people, and will use their best endeavours to facilitate the achievement of this object, it being clearly understood that nothing shall be done which may prejudice the civil and religious rights of non-Jewish communities in Palestine or the rights and political status enjoyed by Jews in any other country **THE BALFOUR DECLARATION 2 NOV 1917**

The British conquered Palestine in 1917-1918, occupying Jerusalem in December 1917

Beirut

Sidon

Litani

Damascus

Tyre

Kuneitra

S Y R I A

Acre

Haifa

Safed

Sea of Galilee

Tiberias

I R A Q

Irbid

River Jordan

Nablus

Salt

Tel Aviv

Jaffa

Ramla

Amman

Ashkelon

Jerusalem

Mediterranean Sea

Gaza

Bethlehem

Dead Sea

Rafah

Kerak

Beersheba

El Arish

Negev

Suez Canal

E G Y P T

Petra

Maan

Eilat

Akaba

H E D J A Z
later SAUDI ARABIA

Sinai

Gulf of Akaba

We Arabs, especially the educated among us, look with deepest sympathy on the Zionist movement.... We will wish the Jews a hearty welcome home.... We are working together for a reformed and revised Near East, and our two movements complement one another. The movement is national and not imperialistic. There is room in Syria for us both. Indeed, I think that neither can be successful without the other **THE EMIR FEISAL TO FELIX FRANKFURTER 3 MARCH 1919**

Makna

If, as may well happen, there should be created in our own lifetime by the banks of the Jordan a Jewish State under the protection of the British Crown which might comprise three or four millions of Jews, an event will have occurred in the history of the world which would from every point of view be beneficial, and would be especially in harmony with the truest interests of the British Empire **WINSTON CHURCHILL ILLUSTRATED SUNDAY HERALD 8 FEB 1920**

Strait of Tiran

Red Sea

© Martin Gilbert

8

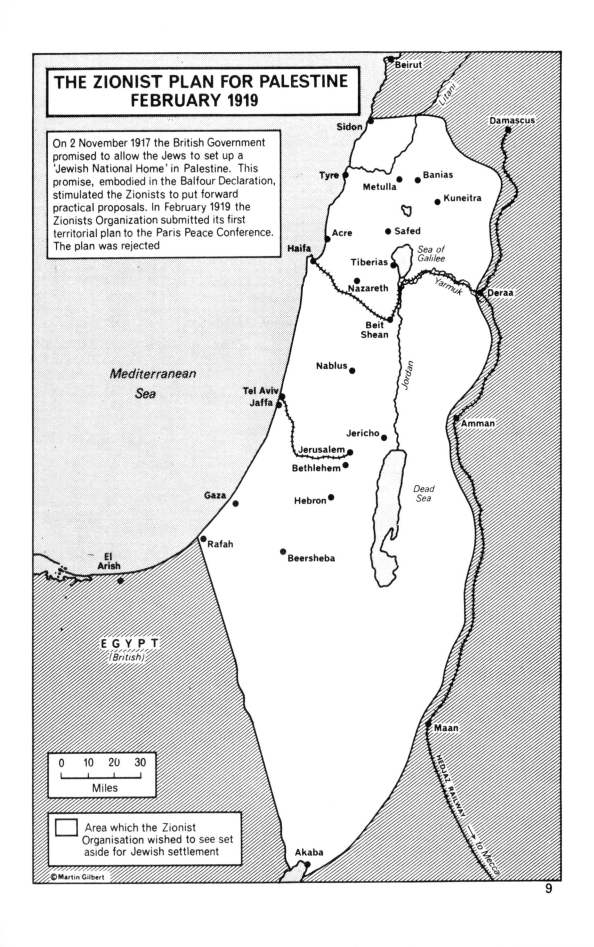

THE ZIONIST PLAN FOR PALESTINE FEBRUARY 1919

On 2 November 1917 the British Government
promised to allow the Jews to set up a
'Jewish National Home' in Palestine. This
promise, embodied in the Balfour Declaration,
stimulated the Zionists to put forward
practical proposals. In February 1919 the
Zionists Organization submitted its first
territorial plan to the Paris Peace Conference.
The plan was rejected

Beirut

Litani

Sidon

Damascus

Tyre

Banias

Metulla

Kuneitra

Acre

Safed

Haifa

Sea of
Galilee

Tiberias

Nazareth

Yarmuk

Deraa

Beit
Shean

Jordan

Nablus

Mediterranean
Sea

Tel Aviv
Jaffa

Jericho

Amman

Jerusalem

Bethlehem

Gaza

Hebron

Dead
Sea

Rafah

El
Arish

Beersheba

EGYPT
(British)

Maan

0	10	20	30

Miles

HEDJAZ RAILWAY ➝ to Mecca

Akaba

☐ Area which the Zionist
Organisation wished to see set
aside for Jewish settlement

© Martin Gilbert

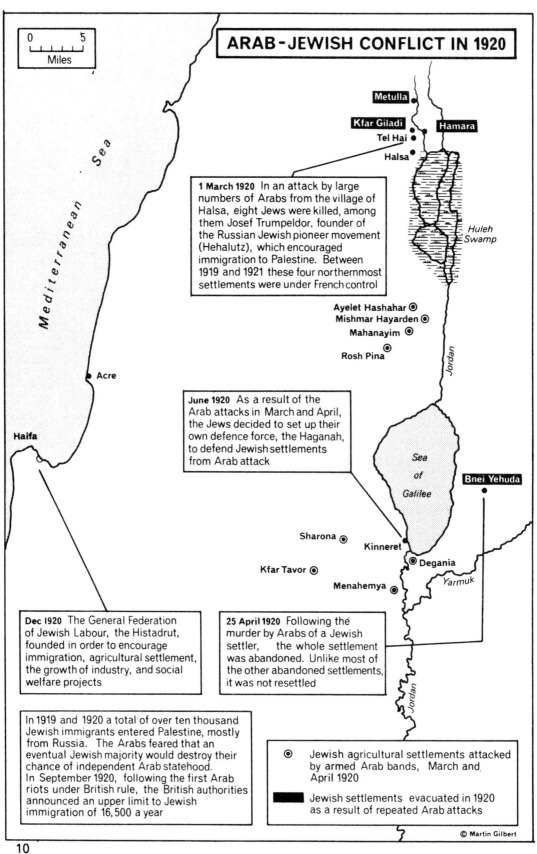

ARAB-JEWISH CONFLICT IN 1920

Metulla

Kfar Giladi **Hamara**
Tel Hai
Halsa

Huleh Swamp

1 March 1920 In an attack by large numbers of Arabs from the village of Halsa, eight Jews were killed, among them Josef Trumpeldor, founder of the Russian Jewish pioneer movement (Hehalutz), which encouraged immigration to Palestine. Between 1919 and 1921 these four northernmost settlements were under French control

Ayelet Hashahar ◉
Mishmar Hayarden ◉
Mahanayim ◉
Rosh Pina ◉

Mediterranean Sea

● **Acre**

Haifa

Jordan

June 1920 As a result of the Arab attacks in March and April, the Jews decided to set up their own defence force, the Haganah, to defend Jewish settlements from Arab attack

Sea of Galilee

Bnei Yehuda ●

Sharona ◉
Kinneret
Kfar Tavor ◉
◉ Degania
Yarmuk
Menahemya ◉

Dec 1920 The General Federation of Jewish Labour, the Histadrut, founded in order to encourage immigration, agricultural settlement, the growth of industry, and social welfare projects

25 April 1920 Following the murder by Arabs of a Jewish settler, the whole settlement was abandoned. Unlike most of the other abandoned settlements, it was not resettled

In 1919 and 1920 a total of over ten thousand Jewish immigrants entered Palestine, mostly from Russia. The Arabs feared that an eventual Jewish majority would destroy their chance of independent Arab statehood. In September 1920, following the first Arab riots under British rule, the British authorities announced an upper limit to Jewish immigration of 16,500 a year

Jordan

◉   Jewish agricultural settlements attacked by armed Arab bands, March and April 1920

▬   Jewish settlements evacuated in 1920 as a result of repeated Arab attacks

© Martin Gilbert

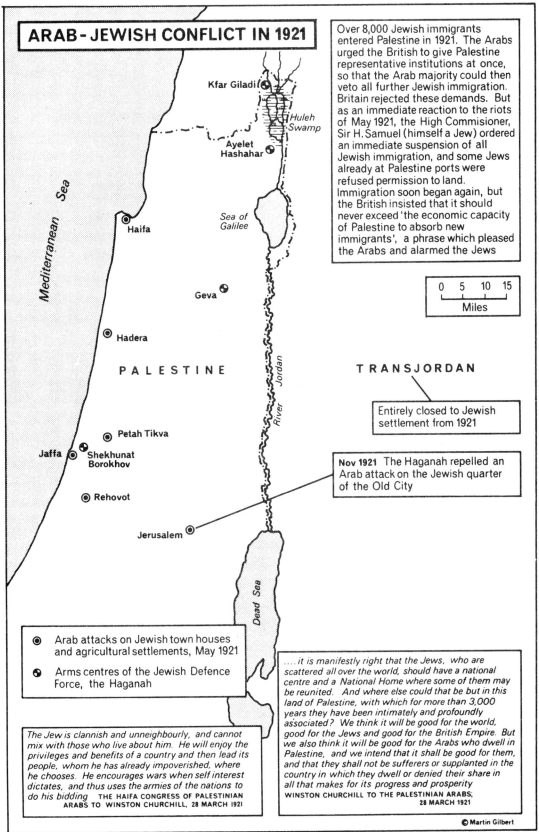

ARAB-JEWISH CONFLICT IN 1921

Over 8,000 Jewish immigrants entered Palestine in 1921. The Arabs urged the British to give Palestine representative institutions at once, so that the Arab majority could then veto all further Jewish immigration. Britain rejected these demands. But as an immediate reaction to the riots of May 1921, the High Commisioner, Sir H. Samuel (himself a Jew) ordered an immediate suspension of all Jewish immigration, and some Jews already at Palestine ports were refused permission to land. Immigration soon began again, but the British insisted that it should never exceed 'the economic capacity of Palestine to absorb new immigrants', a phrase which pleased the Arabs and alarmed the Jews

Kfar Giladi

Huleh Swamp

Ayelet Hashahar

Mediterranean Sea

Sea of Galilee

Haifa

```
0    5    10   15
Miles
```

Geva

Hadera

P A L E S T I N E

River Jordan

T R A N S J O R D A N

Entirely closed to Jewish settlement from 1921

Petah Tikva

Jaffa

Shekhunat Borokhov

Nov 1921 The Haganah repelled an Arab attack on the Jewish quarter of the Old City

Rehovot

Jerusalem

Dead Sea

◉ Arab attacks on Jewish town houses and agricultural settlements, May 1921

◓ Arms centres of the Jewish Defence Force, the Haganah

....it is manifestly right that the Jews, who are scattered all over the world, should have a national centre and a National Home where some of them may be reunited. And where else could that be but in this land of Palestine, with which for more than 3,000 years they have been intimately and profoundly associated? We think it will be good for the world, good for the Jews and good for the British Empire. But we also think it will be good for the Arabs who dwell in Palestine, and we intend that it shall be good for them, and that they shall not be sufferers or supplanted in the country in which they dwell or denied their share in all that makes for its progress and prosperity WINSTON CHURCHILL TO THE PALESTINIAN ARABS, 28 MARCH 1921

The Jew is clannish and unneighbourly, and cannot mix with those who live about him. He will enjoy the privileges and benefits of a country and then lead its people, whom he has already impoverished, where he chooses. He encourages wars when self interest dictates, and thus uses the armies of the nations to do his bidding THE HAIFA CONGRESS OF PALESTINIAN ARABS TO WINSTON CHURCHILL, 28 MARCH 1921

© Martin Gilbert

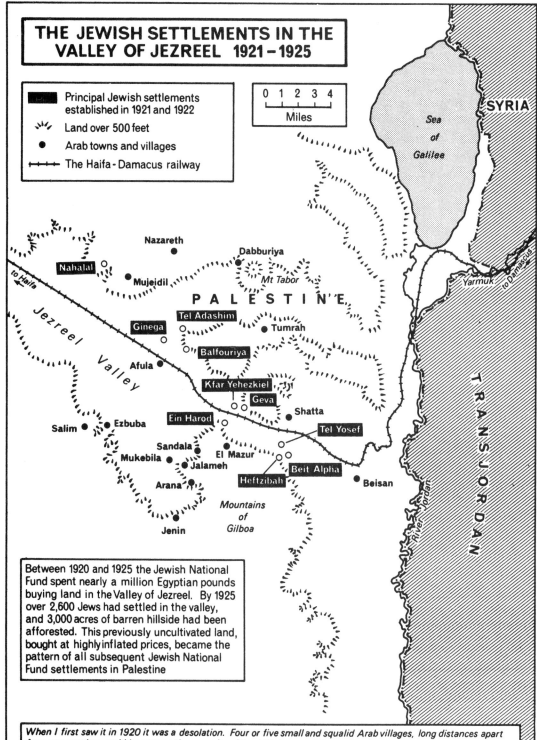

THE JEWISH SETTLEMENTS IN THE VALLEY OF JEZREEL 1921–1925

Principal Jewish settlements established in 1921 and 1922

Land over 500 feet

● Arab towns and villages

The Haifa - Damacus railway

0 1 2 3 4
Miles

SYRIA

Sea of Galilee

to Haifa

Jezreel Valley

Nazareth

Dabburiya

Nahalal

Mujeidil

Mt Tabor

PALESTINE

Tel Adashim

Ginega

Tumrah

Balfouriya

Afula

Kfar Yehezkiel

Geva

Shatta

Ezbuba

Ein Harod

Tel Yosef

Salim

Sandala

El Mazur

Mukebila

Jalameh

Beit Alpha

Arana

Heftzibah

Beisan

Jenin

Mountains of Gilboa

Yarmuk

to Damascus

River Jordan

TRANSJORDAN

Between 1920 and 1925 the Jewish National Fund spent nearly a million Egyptian pounds buying land in the Valley of Jezreel. By 1925 over 2,600 Jews had settled in the valley, and 3,000 acres of barren hillside had been afforested. This previously uncultivated land, bought at highly inflated prices, became the pattern of all subsequent Jewish National Fund settlements in Palestine

When I first saw it in 1920 it was a desolation. Four or five small and squalid Arab villages, long distances apart from one another, could be seen on the summits of low hills here and there. For the rest, the country was uninhabited. There was not a house, not a tree.... about 51 square miles of the valley have now been purchased by the Jewish National Fund..... Twenty schools have been opened. There is an Agricultural Training College for Women in one village and a hospital in another. All the swamps and marshes within the area that has been colonised have been drained.... The whole aspect of the valley has been changed.... in the spring the fields of vegetables or of cereals cover many miles of the land, and what five years ago was little better than a wilderness is being transformed before our eyes into smiling countryside. SIR H. SAMUEL'S REPORT ON THE ADMINISTRATION OF PALESTINE, 22 APRIL 1925

© Martin Gilbert

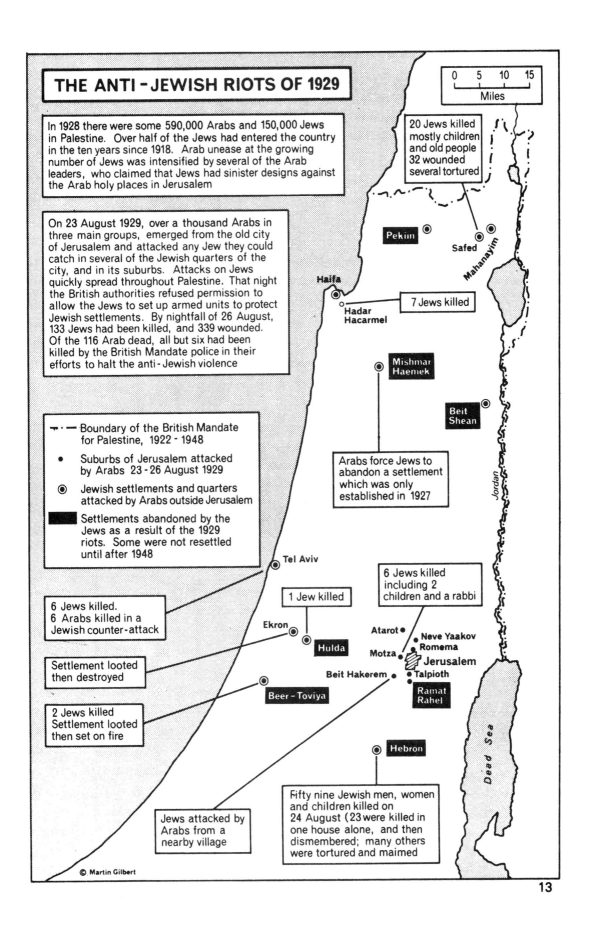

THE ANTI-JEWISH RIOTS OF 1929

0 5 10 15
Miles

In 1928 there were some 590,000 Arabs and 150,000 Jews in Palestine. Over half of the Jews had entered the country in the ten years since 1918. Arab unease at the growing number of Jews was intensified by several of the Arab leaders, who claimed that Jews had sinister designs against the Arab holy places in Jerusalem

On 23 August 1929, over a thousand Arabs in three main groups, emerged from the old city of Jerusalem and attacked any Jew they could catch in several of the Jewish quarters of the city, and in its suburbs. Attacks on Jews quickly spread throughout Palestine. That night the British authorities refused permission to allow the Jews to set up armed units to protect Jewish settlements. By nightfall of 26 August, 133 Jews had been killed, and 339 wounded. Of the 116 Arab dead, all but six had been killed by the British Mandate police in their efforts to halt the anti-Jewish violence

20 Jews killed mostly children and old people 32 wounded several tortured

Pekiin

Safed

Mahanayim

Haifa

7 Jews killed

Hadar Hacarmel

Mishmar Haemek

Beit Shean

Arabs force Jews to abandon a settlement which was only established in 1927

Jordan

—··— Boundary of the British Mandate for Palestine, 1922 - 1948

• Suburbs of Jerusalem attacked by Arabs 23-26 August 1929

◉ Jewish settlements and quarters attacked by Arabs outside Jerusalem

▇ Settlements abandoned by the Jews as a result of the 1929 riots. Some were not resettled until after 1948

Tel Aviv

6 Jews killed including 2 children and a rabbi

1 Jew killed

6 Jews killed.
6 Arabs killed in a Jewish counter-attack

Ekron

Atarot •

Neve Yaakov
Romema

Hulda

Motza •

Jerusalem

Settlement looted then destroyed

Beit Hakerem •

• Talpioth

Beer - Toviya

Ramat Rahel

2 Jews killed
Settlement looted then set on fire

Dead Sea

Hebron

Jews attacked by Arabs from a nearby village

Fifty nine Jewish men, women and children killed on 24 August (23 were killed in one house alone, and then dismembered; many others were tortured and maimed

© Martin Gilbert

13

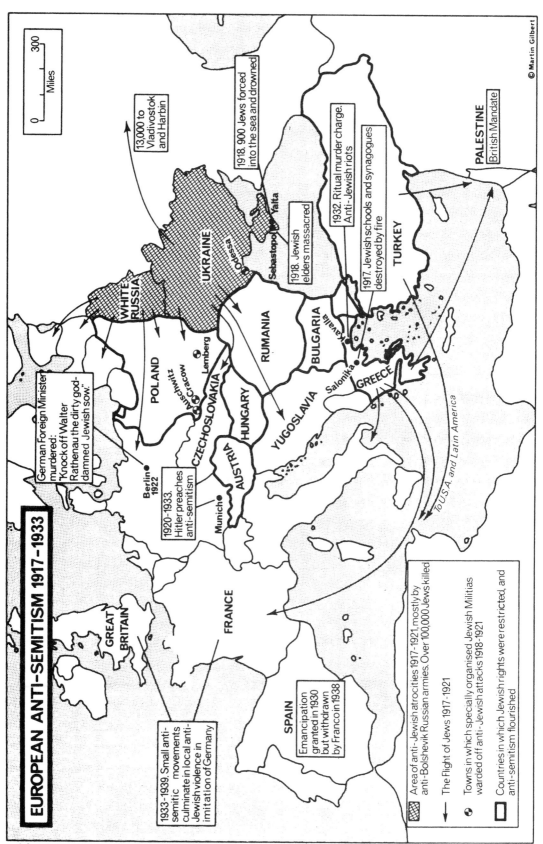

EUROPEAN ANTI-SEMITISM 1917–1933

© Martin Gilbert

300
Miles
0

13,000 to
Vladivostok
and Harbin

1918. 900 Jews forced
into the sea and drowned

1918. Jewish
elders massacred

1932. Ritual murder charge.
Anti-Jewish riots

1917. Jewish schools and synagogues
destroyed by fire

PALESTINE
British Mandate

German Foreign Minister
murdered:
"Knock off Walter
Rathenau the dirty god-
damned Jewish sow."

1920-1933.
Hitler preaches
anti-semitism

1933-1939. Small anti-
semitic movements
culminate in local anti-
Jewish violence in
imitation of Germany

SPAIN
Emancipation
granted in 1930
but withdrawn
by Franco in 1938

GREAT
BRITAIN

FRANCE

POLAND

Auschwitz
Cracow
Lemberg

CZECHOSLOVAKIA

AUSTRIA

HUNGARY

Munich ●

Berlin ●
1922

WHITE
RUSSIA

UKRAINE

Odessa

Sebastopol Yalta

RUMANIA

BULGARIA

YUGOSLAVIA

Kavalla

Salonika ●

GREECE

TURKEY

To U.S.A. and Latin America

Area of anti-Jewish atrocities 1917-1921, mostly by
anti-Bolshevik Russian armies. Over 100,000 Jews killed

The Flight of Jews 1917-1921

Towns in which specially organised Jewish Militias
warded off anti-Jewish attacks 1918-1921

Countries in which Jewish rights were restricted, and
anti-semitism flourished

14

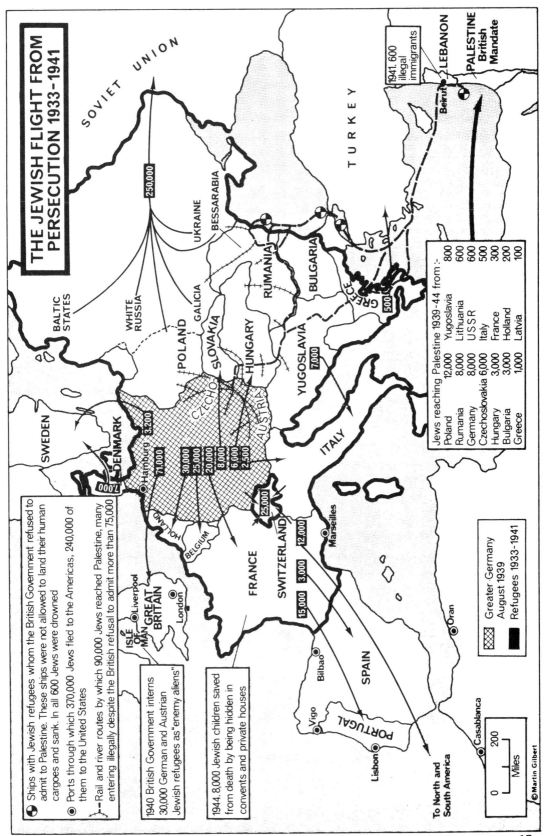

THE JEWISH FLIGHT FROM PERSECUTION 1933–1941

SOVIET UNION

BALTIC STATES

WHITE RUSSIA

UKRAINE

BESSARABIA

250,000

SWEDEN

DENMARK

Hamburg 3,200

7,000

7,000

30,000
25,000
20,000
8,000
6,000
2,500

CZECHO

GALICIA

POLAND

SLOVAKIA

HUNGARY

RUMANIA

BULGARIA

GREECE

500

7,000

AUSTRIA

YUGOSLAVIA

TURKEY

LEBANON

Beirut

PALESTINE British Mandate

1941. 600 illegal immigrants

HOLLAND
BELGIUM

GREAT BRITAIN

ISLE OF MAN

Liverpool

London

25,000

FRANCE

SWITZERLAND

12,000

3,000

15,000

Marseilles

ITALY

SPAIN

Bilbao

Vigo

PORTUGAL

Lisbon

Oran

Casablanca

To North and South America

Jews reaching Palestine 1939–44 from :-

Poland	12,000	Yugoslavia	800
Rumania	8,000	Lithuania	600
Germany	8,000	USSR	600
Czechoslovakia	6,000	Italy	500
Hungary	3,000	France	300
Bulgaria	3,000	Holland	200
Greece	1,000	Latvia	100

▨	Greater Germany August 1939
■	Refugees 1933–1941

Ships with Jewish refugees whom the British Government refused to admit to Palestine. These ships were not allowed to land their human cargoes and sank. In all 600 Jews were drowned

Ports through which 370,000 Jews fled to the Americas, 240,000 of them to the United States

Rail and river routes by which 90,000 Jews reached Palestine, many entering illegally despite the British refusal to admit more than 75,000

1940 British Government interns 30,000 German and Austrian Jewish refugees as "enemy aliens"

1944. 8,000 Jewish children saved from death by being hidden in convents and private houses

0 200

Miles

©Martin Gilbert

ARAB FEARS OF A JEWISH MAJORITY IN PALESTINE
1920 - 1939

TOTAL POPULATION OF PALESTINE

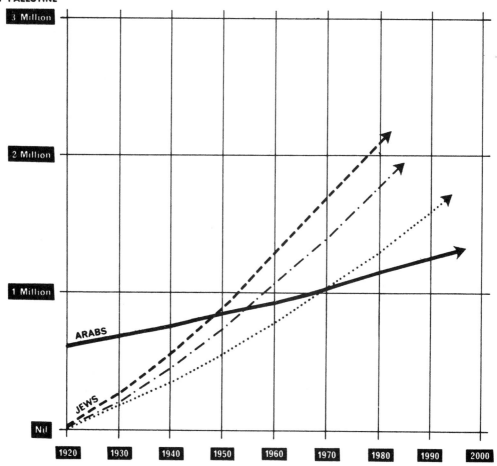

1920	1930	1940	1950	1960	1970	1980	1990	2000

———— Estimated growth of the Arab population of Palestine, 1920-2000

– – – – Estimated Jewish population, allowing for an annual immigration of 25,000 from 1930 (Jews would then equal Arabs by 1948)

– · – · Estimated Jewish population, allowing for an annual immigration restricted to 15,000 (Jews would then equal Arabs by 1956)

·········· Estimated Jewish population, allowing for an annual immigration restricted to 10,000 (Jews would then equal Arabs by 1969)

These estimates were prepared by the British Government in 1929

As well as 360 000 Jewish immigrants between 1919 and 1939, over 50,000 Arabs also immigrated to Palestine (from nearby Arab States) attracted by the improving agricultural conditions and growing job opportunities, most of them created by the Jews

On the 17 May 1939, following a decade of Arab protest, the British Government issued a White Paper restricting Jewish immigration to 15,000 a year for five years, after which no immigration whatsoever would be allowed without Arab permission. The White Paper made it possible for the Arabs to prevent the Jews ever becoming a majority in Palestine

© Martin Gilbert

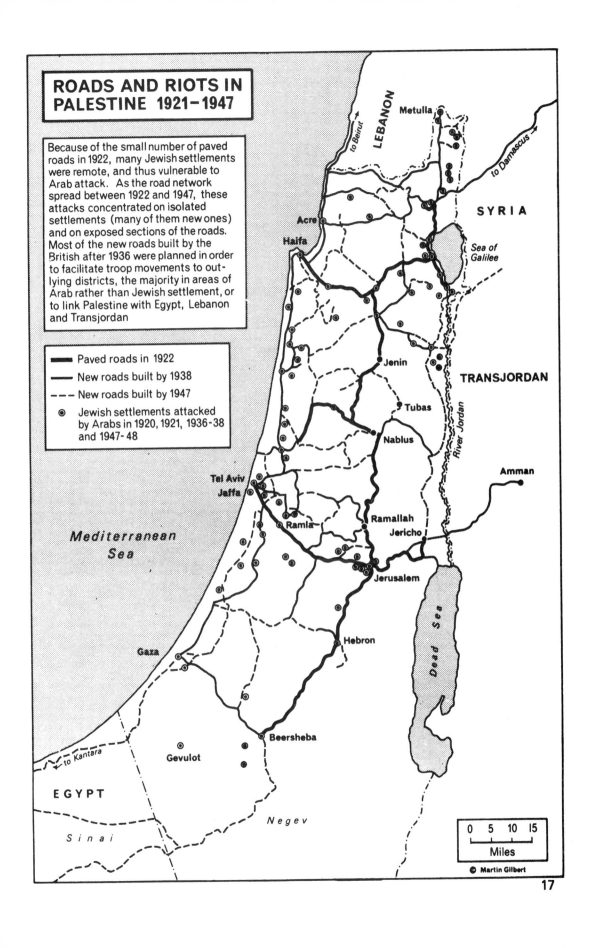

ROADS AND RIOTS IN PALESTINE 1921–1947

Because of the small number of paved roads in 1922, many Jewish settlements were remote, and thus vulnerable to Arab attack. As the road network spread between 1922 and 1947, these attacks concentrated on isolated settlements (many of them new ones) and on exposed sections of the roads. Most of the new roads built by the British after 1936 were planned in order to facilitate troop movements to out-lying districts, the majority in areas of Arab rather than Jewish settlement, or to link Palestine with Egypt, Lebanon and Transjordan

——— Paved roads in 1922

——— New roads built by 1938

- - - New roads built by 1947

◉ Jewish settlements attacked by Arabs in 1920, 1921, 1936-38 and 1947-48

LEBANON

to Beirut

Metulla

to Damascus

SYRIA

Acre

Haifa

Sea of Galilee

Jenin

TRANSJORDAN

Tubas

Nablus

River Jordan

Amman

Tel Aviv

Jaffa

Ramla

Ramallah

Jericho

Jerusalem

Mediterranean Sea

Hebron

Dead Sea

Gaza

to Kantara

Gevulot

Beersheba

EGYPT

Negev

Sinai

0 5 10 15
Miles

© Martin Gilbert

17

THE ARAB CAMPAIGN OF 1936: THE FIRST MONTH

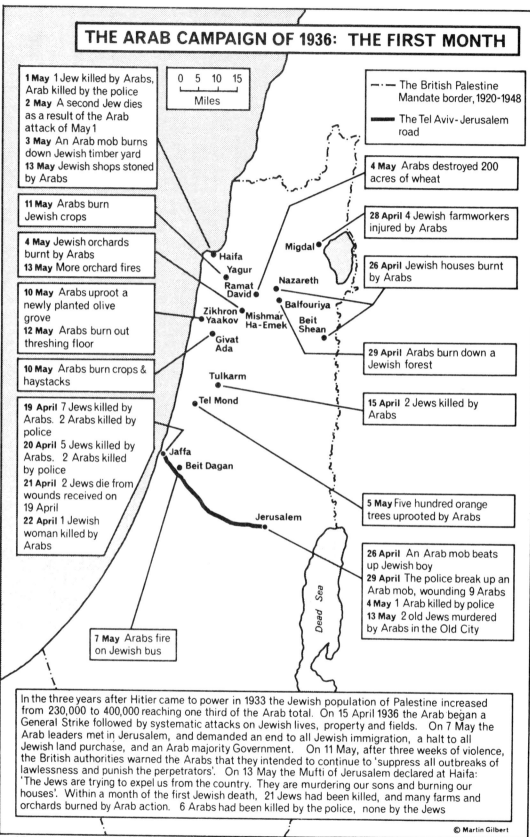

1 May 1 Jew killed by Arabs, Arab killed by the police
2 May A second Jew dies as a result of the Arab attack of May 1
3 May An Arab mob burns down Jewish timber yard
13 May Jewish shops stoned by Arabs

11 May Arabs burn Jewish crops

4 May Jewish orchards burnt by Arabs
13 May More orchard fires

10 May Arabs uproot a newly planted olive grove
12 May Arabs burn out threshing floor

10 May Arabs burn crops & haystacks

19 April 7 Jews killed by Arabs. 2 Arabs killed by police
20 April 5 Jews killed by Arabs. 2 Arabs killed by police
21 April 2 Jews die from wounds received on 19 April
22 April 1 Jewish woman killed by Arabs

7 May Arabs fire on Jewish bus

0 5 10 15
Miles

–·– The British Palestine Mandate border, 1920-1948

▬▬ The Tel Aviv - Jerusalem road

4 May Arabs destroyed 200 acres of wheat

28 April 4 Jewish farmworkers injured by Arabs

26 April Jewish houses burnt by Arabs

29 April Arabs burn down a Jewish forest

15 April 2 Jews killed by Arabs

5 May Five hundred orange trees uprooted by Arabs

26 April An Arab mob beats up Jewish boy
29 April The police break up an Arab mob, wounding 9 Arabs
4 May 1 Arab killed by police
13 May 2 old Jews murdered by Arabs in the Old City

Migdal
Haifa
Yagur
Ramat David
Nazareth
Balfouriya
Zikhron Yaakov
Mishmar Ha-Emek
Beit Shean
Givat Ada
Tulkarm
Tel Mond
Jaffa
Beit Dagan
Jerusalem
Dead Sea

In the three years after Hitler came to power in 1933 the Jewish population of Palestine increased from 230,000 to 400,000 reaching one third of the Arab total. On 15 April 1936 the Arab began a General Strike followed by systematic attacks on Jewish lives, property and fields. On 7 May the Arab leaders met in Jerusalem, and demanded an end to all Jewish immigration, a halt to all Jewish land purchase, and an Arab majority Government. On 11 May, after three weeks of violence, the British authorities warned the Arabs that they intended to continue to 'suppress all outbreaks of lawlessness and punish the perpetrators'. On 13 May the Mufti of Jerusalem declared at Haifa: 'The Jews are trying to expel us from the country. They are murdering our sons and burning our houses'. Within a month of the first Jewish death, 21 Jews had been killed, and many farms and orchards burned by Arab action. 6 Arabs had been killed by the police, none by the Jews

© Martin Gilbert

THE ARAB CAMPAIGN 1936: THE SECOND MONTH

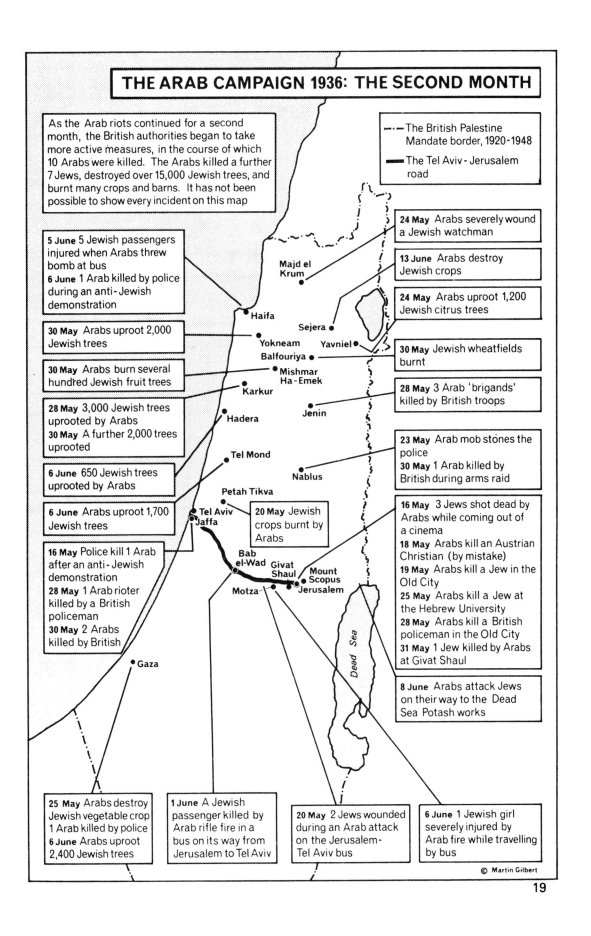

As the Arab riots continued for a second month, the British authorities began to take more active measures, in the course of which 10 Arabs were killed. The Arabs killed a further 7 Jews, destroyed over 15,000 Jewish trees, and burnt many crops and barns. It has not been possible to show every incident on this map

- – · – The British Palestine Mandate border, 1920-1948
- ▬▬ The Tel Aviv - Jerusalem road

5 June 5 Jewish passengers injured when Arabs threw bomb at bus
6 June 1 Arab killed by police during an anti - Jewish demonstration

30 May Arabs uproot 2,000 Jewish trees

30 May Arabs burn several hundred Jewish fruit trees

28 May 3,000 Jewish trees uprooted by Arabs
30 May A further 2,000 trees uprooted

6 June 650 Jewish trees uprooted by Arabs

6 June Arabs uproot 1,700 Jewish trees

16 May Police kill 1 Arab after an anti - Jewish demonstration
28 May 1 Arab rioter killed by a British policeman
30 May 2 Arabs killed by British

24 May Arabs severely wound a Jewish watchman

13 June Arabs destroy Jewish crops

24 May Arabs uproot 1,200 Jewish citrus trees

30 May Jewish wheatfields burnt

28 May 3 Arab 'brigands' killed by British troops

23 May Arab mob stones the police
30 May 1 Arab killed by British during arms raid

16 May 3 Jews shot dead by Arabs while coming out of a cinema
18 May Arabs kill an Austrian Christian (by mistake)
19 May Arabs kill a Jew in the Old City
25 May Arabs kill a Jew at the Hebrew University
28 May Arabs kill a British policeman in the Old City
31 May 1 Jew killed by Arabs at Givat Shaul

8 June Arabs attack Jews on their way to the Dead Sea Potash works

20 May Jewish crops burnt by Arabs

Majd el Krum

Haifa

Sejera

Yokneam   Yavniel

Balfouriya

Mishmar Ha - Emek

Karkur

Hadera

Jenin

Nablus

Tel Mond

Petah Tikva

Tel Aviv
Jaffa

Bab el-Wad   Givat Shaul

Motza   Mount Scopus
Jerusalem

Gaza

Dead Sea

25 May Arabs destroy Jewish vegetable crop 1 Arab killed by police
6 June Arabs uproot 2,400 Jewish trees

1 June A Jewish passenger killed by Arab rifle fire in a bus on its way from Jerusalem to Tel Aviv

20 May 2 Jews wounded during an Arab attack on the Jerusalem-Tel Aviv bus

6 June 1 Jewish girl severely injured by Arab fire while travelling by bus

© Martin Gilbert

19

THE ARAB CAMPAIGN OF 1936: THE THIRD MONTH

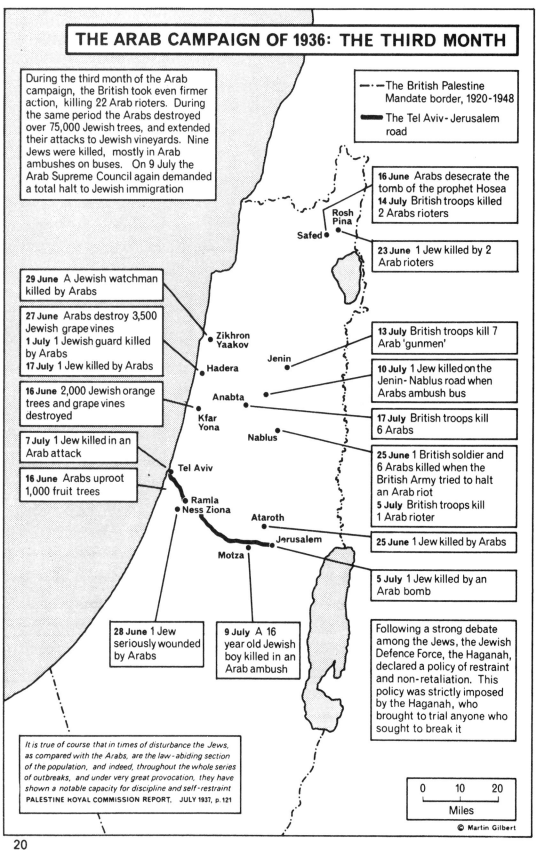

During the third month of the Arab campaign, the British took even firmer action, killing 22 Arab rioters. During the same period the Arabs destroyed over 75,000 Jewish trees, and extended their attacks to Jewish vineyards. Nine Jews were killed, mostly in Arab ambushes on buses. On 9 July the Arab Supreme Council again demanded a total halt to Jewish immigration

---·—· The British Palestine Mandate border, 1920-1948

━━━ The Tel Aviv - Jerusalem road

16 June Arabs desecrate the tomb of the prophet Hosea
14 July British troops killed 2 Arabs rioters

23 June 1 Jew killed by 2 Arab rioters

13 July British troops kill 7 Arab 'gunmen'

10 July 1 Jew killed on the Jenin- Nablus road when Arabs ambush bus

17 July British troops kill 6 Arabs

25 June 1 British soldier and 6 Arabs killed when the British Army tried to halt an Arab riot
5 July British troops kill 1 Arab rioter

25 June 1 Jew killed by Arabs

5 July 1 Jew killed by an Arab bomb

29 June A Jewish watchman killed by Arabs

27 June Arabs destroy 3,500 Jewish grape vines
1 July 1 Jewish guard killed by Arabs
17 July 1 Jew killed by Arabs

16 June 2,000 Jewish orange trees and grape vines destroyed

7 July 1 Jew killed in an Arab attack

16 June Arabs uproot 1,000 fruit trees

28 June 1 Jew seriously wounded by Arabs

9 July A 16 year old Jewish boy killed in an Arab ambush

Following a strong debate among the Jews, the Jewish Defence Force, the Haganah, declared a policy of restraint and non-retaliation. This policy was strictly imposed by the Haganah, who brought to trial anyone who sought to break it

Rosh Pina

Safed

Zikhron Yaakov

Jenin

Hadera

Anabta

Kfar Yona

Nablus

Tel Aviv

Ramla
Ness Ziona

Ataroth

Jerusalem

Motza

It is true of course that in times of disturbance the Jews, as compared with the Arabs, are the law-abiding section of the population, and indeed, throughout the whole series of outbreaks, and under very great provocation, they have shown a notable capacity for discipline and self-restraint
PALESTINE ROYAL COMMISSION REPORT,   JULY 1937, p. 121

0   10   20
Miles

© Martin Gilbert

20

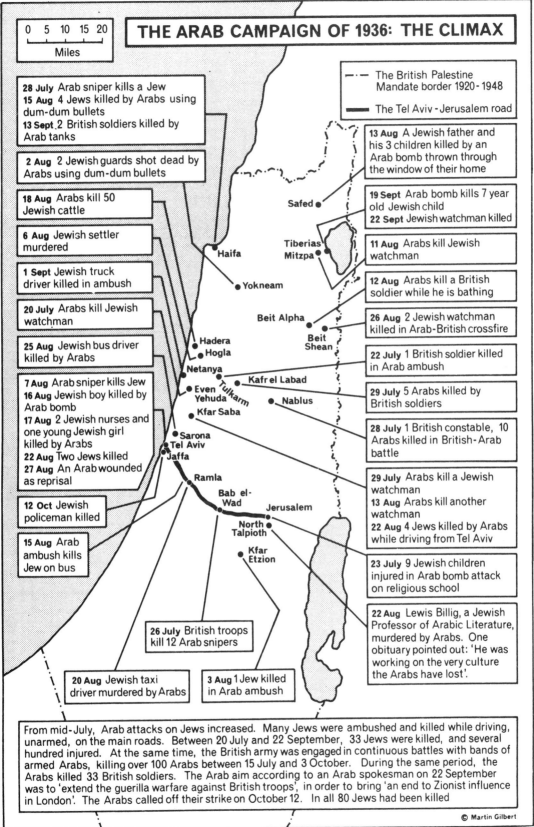

THE ARAB CAMPAIGN OF 1936: THE CLIMAX

0 5 10 15 20
Miles

—·— The British Palestine Mandate border 1920-1948

—— The Tel Aviv - Jerusalem road

28 July Arab sniper kills a Jew
15 Aug 4 Jews killed by Arabs using dum-dum bullets
13 Sept 2 British soldiers killed by Arab tanks

2 Aug 2 Jewish guards shot dead by Arabs using dum-dum bullets

18 Aug Arabs kill 50 Jewish cattle

6 Aug Jewish settler murdered

1 Sept Jewish truck driver killed in ambush

20 July Arabs kill Jewish watchman

25 Aug Jewish bus driver killed by Arabs

7 Aug Arab sniper kills Jew
16 Aug Jewish boy killed by Arab bomb
17 Aug 2 Jewish nurses and one young Jewish girl killed by Arabs
22 Aug Two Jews killed
27 Aug An Arab wounded as reprisal

12 Oct Jewish policeman killed

15 Aug Arab ambush kills Jew on bus

13 Aug A Jewish father and his 3 children killed by an Arab bomb thrown through the window of their home

19 Sept Arab bomb kills 7 year old Jewish child
22 Sept Jewish watchman killed

11 Aug Arabs kill Jewish watchman

12 Aug Arabs kill a British soldier while he is bathing

26 Aug 2 Jewish watchman killed in Arab-British crossfire

22 July 1 British soldier killed in Arab ambush

29 July 5 Arabs killed by British soldiers

28 July 1 British constable, 10 Arabs killed in British-Arab battle

29 July Arabs kill a Jewish watchman
13 Aug Arabs kill another watchman
22 Aug 4 Jews killed by Arabs while driving from Tel Aviv

23 July 9 Jewish children injured in Arab bomb attack on religious school

22 Aug Lewis Billig, a Jewish Professor of Arabic Literature, murdered by Arabs. One obituary pointed out: 'He was working on the very culture the Arabs have lost'.

26 July British troops kill 12 Arab snipers

20 Aug Jewish taxi driver murdered by Arabs

3 Aug 1 Jew killed in Arab ambush

Safed
Haifa
Tiberias
Mitzpa
Yokneam
Beit Alpha
Beit Shean
Hadera
Hogla
Netanya
Kafr el Labad
Even Yehuda
Tulkarm
Nablus
Kfar Saba
Sarona
Tel Aviv
Jaffa
Ramla
Bab el-Wad
Jerusalem
North Talpioth
Kfar Etzion

From mid-July, Arab attacks on Jews increased. Many Jews were ambushed and killed while driving, unarmed, on the main roads. Between 20 July and 22 September, 33 Jews were killed, and several hundred injured. At the same time, the British army was engaged in continuous battles with bands of armed Arabs, killing over 100 Arabs between 15 July and 3 October. During the same period, the Arabs killed 33 British soldiers. The Arab aim according to an Arab spokesman on 22 September was to 'extend the guerilla warfare against British troops', in order to bring 'an end to Zionist influence in London'. The Arabs called off their strike on October 12. In all 80 Jews had been killed

© Martin Gilbert

21

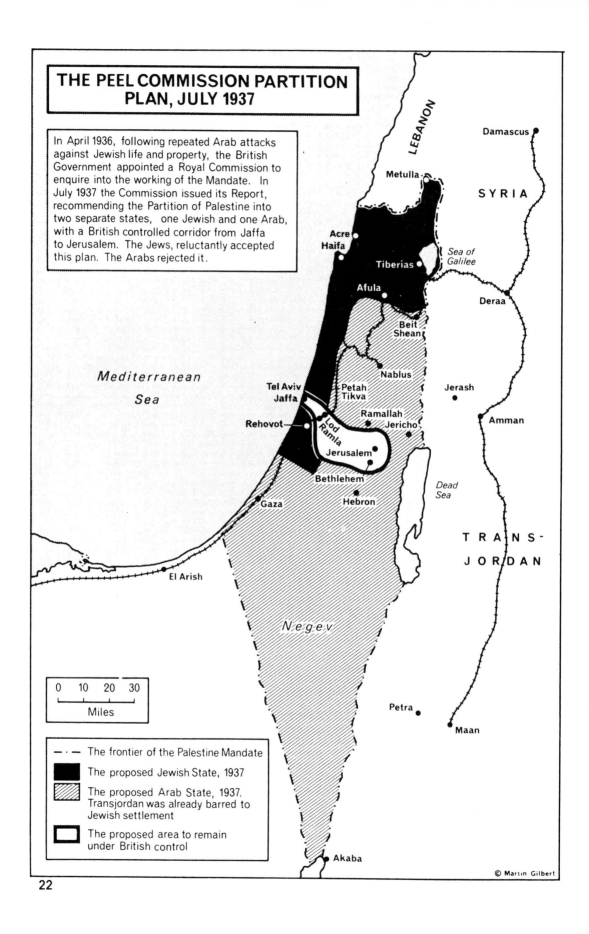

THE PEEL COMMISSION PARTITION PLAN, JULY 1937

In April 1936, following repeated Arab attacks against Jewish life and property, the British Government appointed a Royal Commission to enquire into the working of the Mandate. In July 1937 the Commission issued its Report, recommending the Partition of Palestine into two separate states, one Jewish and one Arab, with a British controlled corridor from Jaffa to Jerusalem. The Jews, reluctantly accepted this plan. The Arabs rejected it.

LEBANON

Damascus

Metulla

SYRIA

Acre
Haifa

Sea of
Galilee

Tiberias

Afula

Deraa

Beit
Shean

Mediterranean

Sea

Nablus

Jerash

Tel Aviv
Jaffa

Petah
Tikva

Ramallah

Amman

Rehovot

Lod
Ramla

Jericho

Jerusalem

Bethlehem

Dead
Sea

Gaza

Hebron

T R A N S -

El Arish

J O R D A N

N e g e v

| 0 | 10 | 20 | 30 |
Miles

Petra

Maan

— · — The frontier of the Palestine Mandate

■ The proposed Jewish State, 1937

▨ The proposed Arab State, 1937.
Transjordan was already barred to
Jewish settlement

▢ The proposed area to remain
under British control

Akaba

© Martin Gilbert

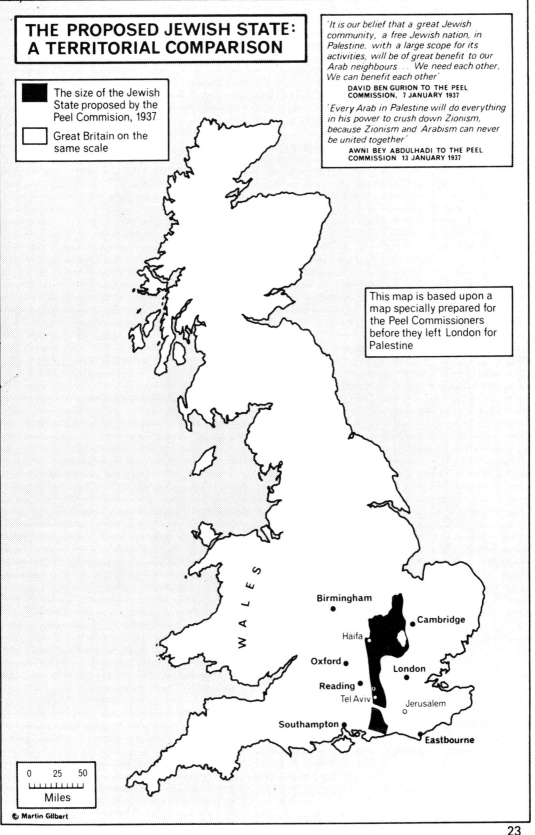

THE PROPOSED JEWISH STATE:
A TERRITORIAL COMPARISON

■ The size of the Jewish State proposed by the Peel Commision, 1937

☐ Great Britain on the same scale

'It is our belief that a great Jewish community, a free Jewish nation, in Palestine, with a large scope for its activities, will be of great benefit to our Arab neighbours... We need each other, We can benefit each other'

DAVID BEN GURION TO THE PEEL COMMISSION, 7 JANUARY 1937

'Every Arab in Palestine will do everything in his power to crush down Zionism, because Zionism and Arabism can never be united together'

AWNI BEY ABDULHADI TO THE PEEL COMMISSION 13 JANUARY 1937

This map is based upon a map specially prepared for the Peel Commissioners before they left London for Palestine

WALES

Birmingham

Cambridge

Haifa

Oxford

London

Reading

Tel Aviv

Jerusalem

Southampton

Eastbourne

0   25   50
Miles

© Martin Gilbert

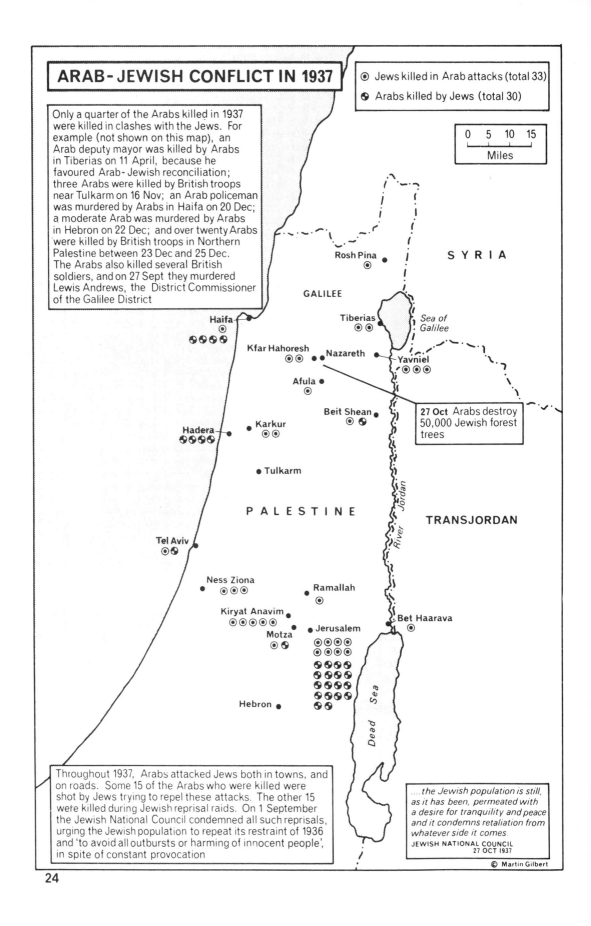

ARAB-JEWISH CONFLICT IN 1937

Legend:
- ⊙ Jews killed in Arab attacks (total 33)
- ⊕ Arabs killed by Jews (total 30)

Scale: 0 5 10 15 Miles

Only a quarter of the Arabs killed in 1937 were killed in clashes with the Jews. For example (not shown on this map), an Arab deputy mayor was killed by Arabs in Tiberias on 11 April, because he favoured Arab-Jewish reconciliation; three Arabs were killed by British troops near Tulkarm on 16 Nov; an Arab policeman was murdered by Arabs in Haifa on 20 Dec; a moderate Arab was murdered by Arabs in Hebron on 22 Dec; and over twenty Arabs were killed by British troops in Northern Palestine between 23 Dec and 25 Dec. The Arabs also killed several British soldiers, and on 27 Sept they murdered Lewis Andrews, the District Commissioner of the Galilee District

SYRIA

Rosh Pina

GALILEE

Tiberias

Sea of Galilee

Haifa

Kfar Hahoresh

Nazareth

Yavniel

Afula

Beit Shean

27 Oct Arabs destroy 50,000 Jewish forest trees

Karkur

Hadera

Tulkarm

PALESTINE

River Jordan

TRANSJORDAN

Tel Aviv

Ness Ziona

Ramallah

Kiryat Anavim

Bet Haarava

Motza

Jerusalem

Dead Sea

Hebron

Throughout 1937, Arabs attacked Jews both in towns, and on roads. Some 15 of the Arabs who were killed were shot by Jews trying to repel these attacks. The other 15 were killed during Jewish reprisal raids. On 1 September the Jewish National Council condemned all such reprisals, urging the Jewish population to repeat its restraint of 1936 and 'to avoid all outbursts or harming of innocent people', in spite of constant provocation

....the Jewish population is still, as it has been, permeated with a desire for tranquility and peace and it condemns retaliation from whatever side it comes.
JEWISH NATIONAL COUNCIL
27 OCT 1937

© Martin Gilbert

24

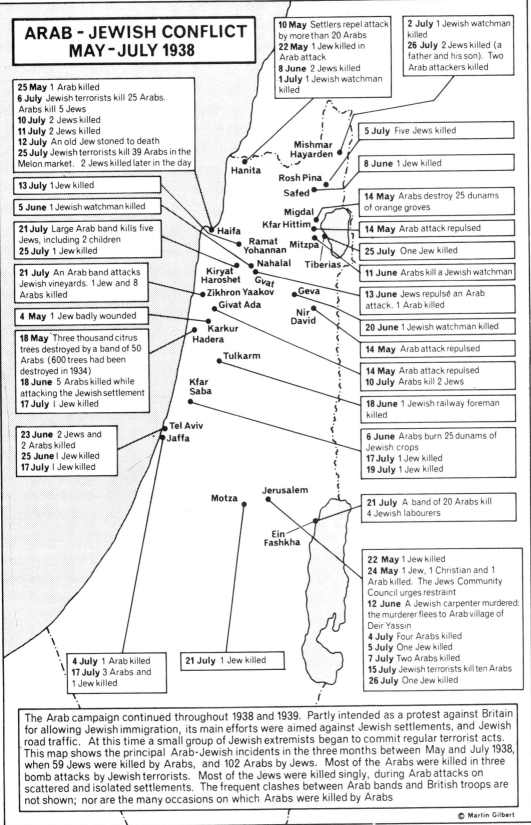

ARAB - JEWISH CONFLICT MAY - JULY 1938

10 May Settlers repel attack by more than 20 Arabs
22 May 1 Jew killed in Arab attack
8 June 2 Jews killed
1 July 1 Jewish watchman killed

2 July 1 Jewish watchman killed
26 July 2 Jews killed (a father and his son). Two Arab attackers killed

25 May 1 Arab killed
6 July Jewish terrorists kill 25 Arabs. Arabs kill 5 Jews
10 July 2 Jews killed
11 July 2 Jews killed
12 July An old Jew stoned to death
25 July Jewish terrorists kill 39 Arabs in the Melon market. 2 Jews killed later in the day

5 July Five Jews killed

8 June 1 Jew killed

13 July 1 Jew killed

5 June 1 Jewish watchman killed

21 July Large Arab band kills five Jews, including 2 children
25 July 1 Jew killed

21 July An Arab band attacks Jewish vineyards. 1 Jew and 8 Arabs killed

4 May 1 Jew badly wounded

18 May Three thousand citrus trees destroyed by a band of 50 Arabs (600 trees had been destroyed in 1934)
18 June 5 Arabs killed while attacking the Jewish settlement
17 July 1 Jew killed

23 June 2 Jews and 2 Arabs killed
25 June 1 Jew killed
17 July 1 Jew killed

14 May Arabs destroy 25 dunams of orange groves

14 May Arab attack repulsed

25 July One Jew killed

11 June Arabs kill a Jewish watchman

13 June Jews repulse an Arab attack. 1 Arab killed

20 June 1 Jewish watchman killed

14 May Arab attack repulsed

14 May Arab attack repulsed
10 July Arabs kill 2 Jews

18 June 1 Jewish railway foreman killed

6 June Arabs burn 25 dunams of Jewish crops
17 July 1 Jew killed
19 July 1 Jew killed

21 July A band of 20 Arabs kill 4 Jewish labourers

22 May 1 Jew killed
24 May 1 Jew, 1 Christian and 1 Arab killed. The Jews Community Council urges restraint
12 June A Jewish carpenter murdered: the murderer flees to Arab village of Deir Yassin
4 July Four Arabs killed
5 July One Jew killed
7 July Two Arabs killed
15 July Jewish terrorists kill ten Arabs
26 July One Jew killed

4 July 1 Arab killed
17 July 3 Arabs and 1 Jew killed

21 July 1 Jew killed

Mishmar Hayarden
Hanita
Rosh Pina
Safed
Migdal
Kfar Hittim
Haifa
Ramat Yohannan
Mitzpa
Nahalal
Tiberias
Kiryat Haroshet
Gvat
Zikhron Yaakov
Geva
Givat Ada
Nir David
Karkur
Hadera
Tulkarm
Kfar Saba
Tel Aviv
Jaffa
Jerusalem
Motza
Ein Fashkha

The Arab campaign continued throughout 1938 and 1939. Partly intended as a protest against Britain for allowing Jewish immigration, its main efforts were aimed against Jewish settlements, and Jewish road traffic. At this time a small group of Jewish extremists began to commit regular terrorist acts. This map shows the principal Arab-Jewish incidents in the three months between May and July 1938, when 59 Jews were killed by Arabs, and 102 Arabs by Jews. Most of the Arabs were killed in three bomb attacks by Jewish terrorists. Most of the Jews were killed singly, during Arab attacks on scattered and isolated settlements. The frequent clashes between Arab bands and British troops are not shown; nor are the many occasions on which Arabs were killed by Arabs

© Martin Gilbert

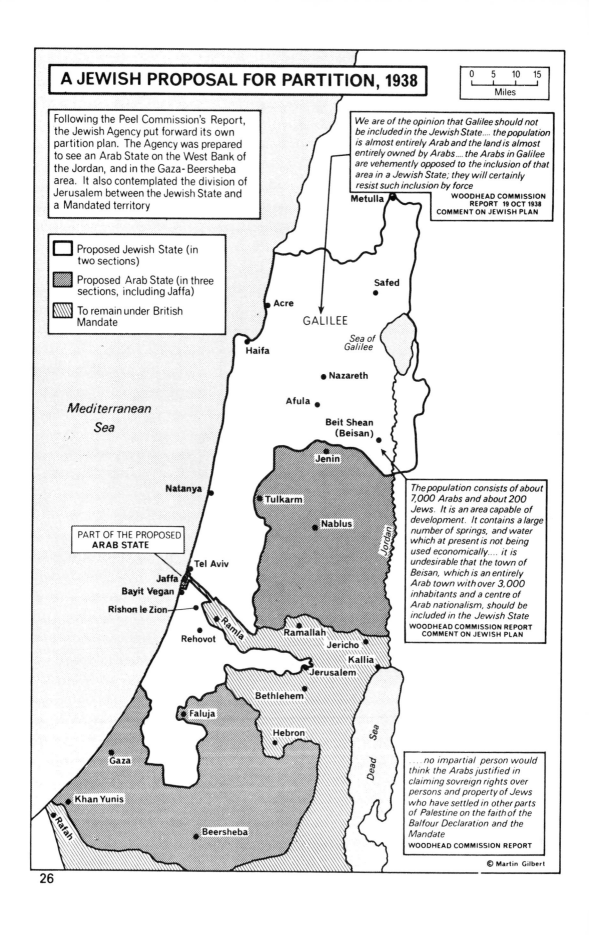

A JEWISH PROPOSAL FOR PARTITION, 1938

0 5 10 15
Miles

Following the Peel Commission's Report, the Jewish Agency put forward its own partition plan. The Agency was prepared to see an Arab State on the West Bank of the Jordan, and in the Gaza-Beersheba area. It also contemplated the division of Jerusalem between the Jewish State and a Mandated territory

We are of the opinion that Galilee should not be included in the Jewish State.... the population is almost entirely Arab and the land is almost entirely owned by Arabs.... the Arabs in Galilee are vehemently opposed to the inclusion of that area in a Jewish State; they will certainly resist such inclusion by force
WOODHEAD COMMISSION REPORT 19 OCT 1938 COMMENT ON JEWISH PLAN

☐ Proposed Jewish State (in two sections)

▨ Proposed Arab State (in three sections, including Jaffa)

▧ To remain under British Mandate

Metulla

Safed

Acre

GALILEE

Sea of Galilee

Haifa

Mediterranean Sea

Nazareth

Afula

Beit Shean (Beisan)

Jenin

The population consists of about 7,000 Arabs and about 200 Jews. It is an area capable of development. It contains a large number of springs, and water which at present is not being used economically.... it is undesirable that the town of Beisan, which is an entirely Arab town with over 3,000 inhabitants and a centre of Arab nationalism, should be included in the Jewish State
WOODHEAD COMMISSION REPORT COMMENT ON JEWISH PLAN

Natanya

Tulkarm

Nablus

Jordan

PART OF THE PROPOSED **ARAB STATE**

Tel Aviv

Jaffa

Bayit Vegan

Rishon le Zion

Ramla

Rehovot

Ramallah

Jericho

Kallia

Jerusalem

Bethlehem

Faluja

Hebron

Dead Sea

Gaza

....no impartial person would think the Arabs justified in claiming sovereign rights over persons and property of Jews who have settled in other parts of Palestine on the faith of the Balfour Declaration and the Mandate
WOODHEAD COMMISSION REPORT

Khan Yunis

Rafah

Beersheba

© Martin Gilbert

26

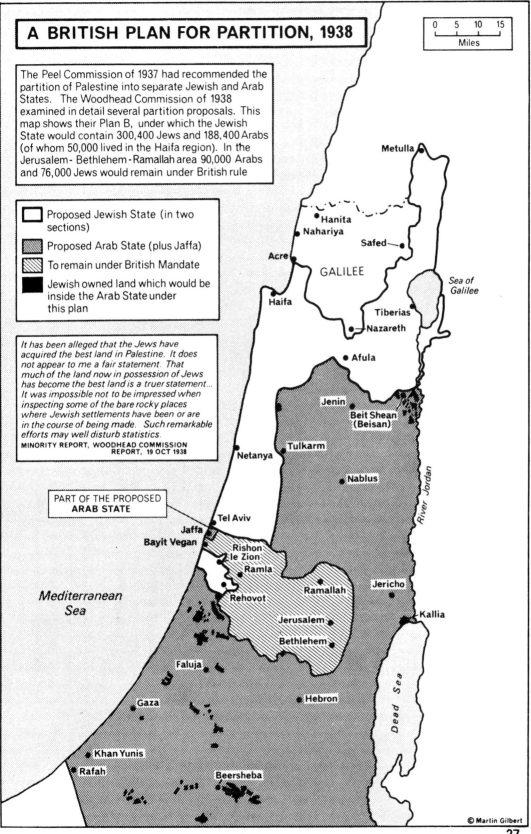

A BRITISH PLAN FOR PARTITION, 1938

0 5 10 15
Miles

The Peel Commission of 1937 had recommended the partition of Palestine into separate Jewish and Arab States. The Woodhead Commission of 1938 examined in detail several partition proposals. This map shows their Plan B, under which the Jewish State would contain 300,400 Jews and 188,400 Arabs (of whom 50,000 lived in the Haifa region). In the Jerusalem-Bethlehem-Ramallah area 90,000 Arabs and 76,000 Jews would remain under British rule

Proposed Jewish State (in two sections)

Proposed Arab State (plus Jaffa)

To remain under British Mandate

Jewish owned land which would be inside the Arab State under this plan

It has been alleged that the Jews have acquired the best land in Palestine. It does not appear to me a fair statement. That much of the land now in possession of Jews has become the best land is a truer statement... It was impossible not to be impressed when inspecting some of the bare rocky places where Jewish settlements have been or are in the course of being made. Such remarkable efforts may well disturb statistics.
MINORITY REPORT, WOODHEAD COMMISSION REPORT, 19 OCT 1938

Metulla

Hanita

Nahariya

Safed

Acre

GALILEE

Sea of Galilee

Haifa

Tiberias

Nazareth

Afula

Jenin

Beit Shean (Beisan)

Tulkarm

Netanya

Nablus

River Jordan

PART OF THE PROPOSED **ARAB STATE**

Tel Aviv

Jaffa

Bayit Vegan

Rishon le Zion

Ramla

Ramallah

Jericho

Rehovot

Kallia

Mediterranean Sea

Jerusalem

Bethlehem

Faluja

Gaza

Hebron

Dead Sea

Khan Yunis

Rafah

Beersheba

© Martin Gilbert

27

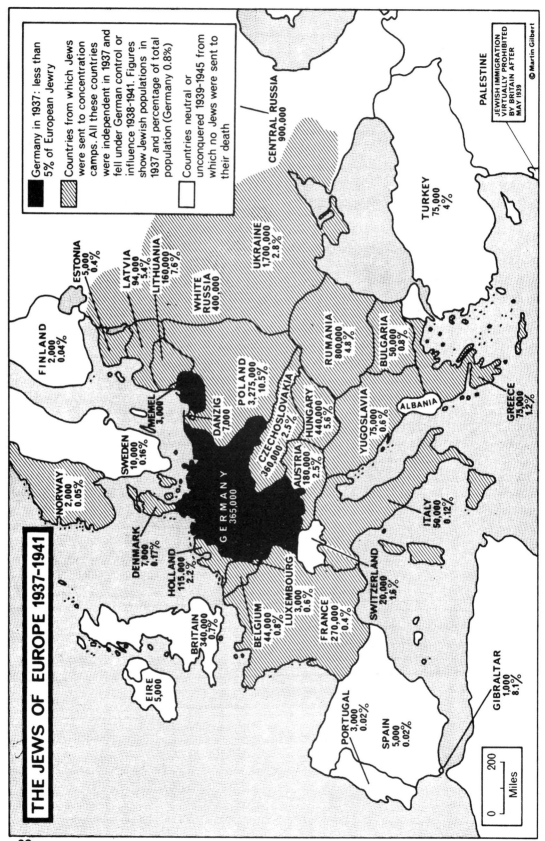

THE JEWS OF EUROPE 1937-1941

Germany in 1937: less than 5% of European Jewry

Countries from which Jews were sent to concentration camps. All these countries were independent in 1937 and fell under German control or influence 1938-1941. Figures show Jewish populations in 1937 and percentage of total population (Germany 0.8%)

Countries neutral or unconquered 1939-1945 from which no Jews were sent to their death

CENTRAL RUSSIA
900,000

PALESTINE

JEWISH IMMIGRATION
VIRTUALLY PROHIBITED
BY BRITAIN AFTER
MAY 1939

© Martin Gilbert

TURKEY
75,000
4%

ESTONIA
5,000
0.4%

LATVIA
94,000
5.4%

LITHUANIA
160,000
7.6%

WHITE RUSSIA
400,000

UKRAINE
1,700,000
2.8%

FINLAND
2,000
0.04%

MEMEL
3,000

DANZIG
7,000

POLAND
3,275,000
10.5%

RUMANIA
800,000
4.8%

BULGARIA
50,000
0.8%

SWEDEN
10,000
0.16%

CZECHOSLOVAKIA
360,000
2.5%

HUNGARY
440,000
5.6%

YUGOSLAVIA
75,000
0.6%

ALBANIA

GREECE
75,000
1.2%

NORWAY
2,000
0.05%

GERMANY
365,000

AUSTRIA
180,000
2.5%

DENMARK
7,000
0.17%

HOLLAND
115,000
2.2%

BELGIUM
44,000
0.8%

LUXEMBOURG
3,000
0.6%

FRANCE
270,000
0.4%

SWITZERLAND
20,000
1.6%

ITALY
50,000
0.12%

BRITAIN
340,000
0.7%

EIRE
5,000

PORTUGAL
3,000
0.02%

SPAIN
5,000
0.02%

GIBRALTAR
1,000
8.1%

0   200
Miles

28

THE VOYAGE OF THE "ST.LOUIS" MAY-JUNE 1939
THE JEWISH SEARCH FOR REFUGE

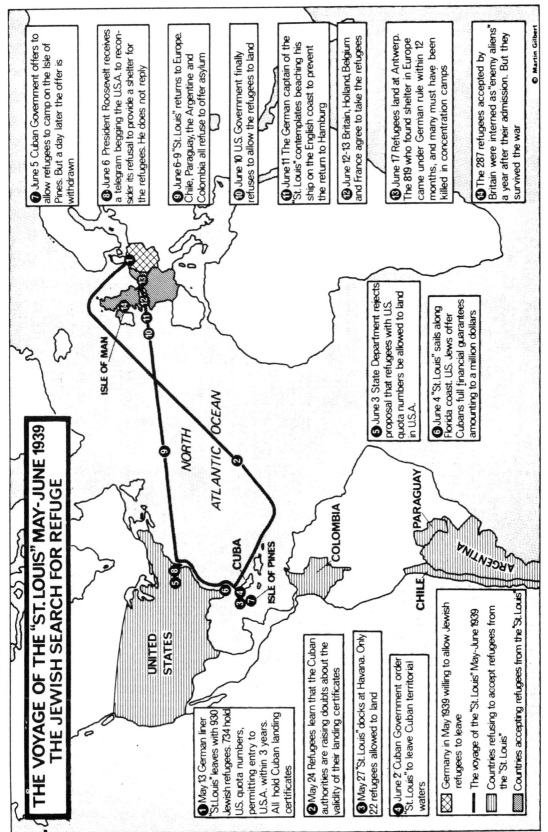

1 May 13 German liner "St.Louis" leaves with 930 Jewish refugees. 734 hold U.S. quota numbers, permitting entry to U.S.A. within 3 years. All hold Cuban landing certificates

2 May 24 Refugees learn that the Cuban authorities are raising doubts about the validity of their landing certificates

3 May 27 "St.Louis" docks at Havana. Only 22 refugees allowed to land

4 June 2 Cuban Government order "St.Louis" to leave Cuban territorial waters

5 June 3 State Deparfment rejects proposal that refugees with U.S. quota numbers be allowed to land in U.S.A.

6 June 4 "St.Louis" sails along Florida coast. U.S. Jews offer Cubans full financial guarantees amounting to a million dollars

7 June 5 Cuban Government offers to allow refugees to camp on the Isle of Pines. But a day later the offer is withdrawn

8 June 6 President Roosevelt receives a telegram begging the U.S.A. to reconsider its refusal to provide a shelter for the refugees. He does not reply

9 June 6-9 "St.Louis" returns to Europe. Chile, Paraguay, the Argentine and Colombia all refuse to offer asylum

10 June 10 U.S. Government finally refuses to allow the refugees to land

11 June 11 The German captain of the "St.Louis" contemplates beaching his ship on the English coast to prevent the return to Hamburg

12 June 12-13 Britain, Holland, Belgium and France agree to take the refugees

13 June 17 Refugees land at Antwerp. The 819 who found shelter in Europe came under German rule within 12 months, and many must have been killed in concentration camps

14 The 287 refugees accepted by Britain were interned as "enemy aliens" a year after their admission. But they survived the war

© Martin Gilbert

⊠ Germany in May 1939 willing to allow Jewish refugees to leave

— The voyage of the "St. Louis" May–June 1939

⦙ Countries refusing to accept refugees from the "St. Louis"

⧗ Countries accepting refugees from the "St. Louis"

NORTH ATLANTIC OCEAN

ISLE OF MAN

UNITED STATES

CUBA

ISLE OF PINES

COLOMBIA

PARAGUAY

CHILE

ARGENTINA

THE SEARCH FOR SAFETY 1933-1945

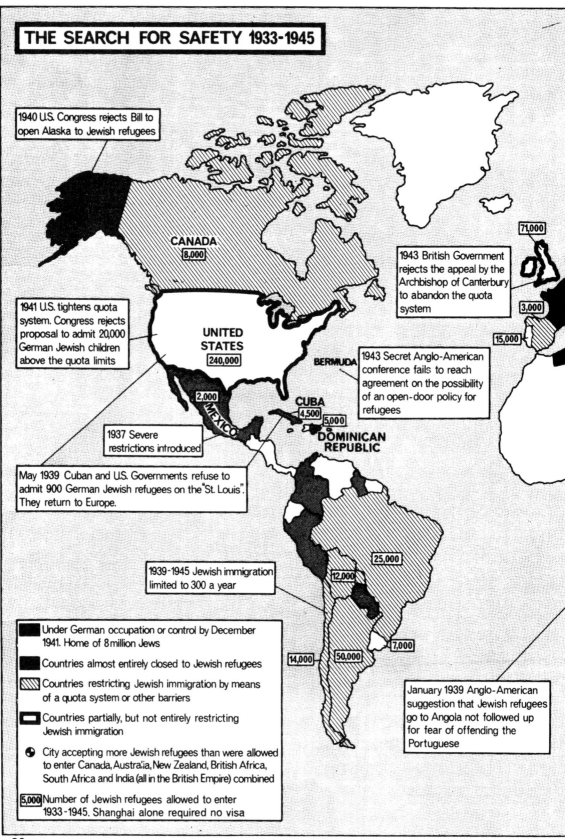

1940 U.S. Congress rejects Bill to open Alaska to Jewish refugees

1943 British Government rejects the appeal by the Archbishop of Canterbury to abandon the quota system

71,000

3,000

15,000

1941 U.S. tightens quota system. Congress rejects proposal to admit 20,000 German Jewish children above the quota limits

CANADA
8,000

UNITED STATES
240,000

BERMUDA

1943 Secret Anglo-American conference fails to reach agreement on the possibility of an open-door policy for refugees

2,000

MEXICO

CUBA
4,500 **5,000**

DOMINICAN REPUBLIC

1937 Severe restrictions introduced

May 1939 Cuban and U.S. Governments refuse to admit 900 German Jewish refugees on the "St. Louis". They return to Europe.

25,000

12,000

1939-1945 Jewish immigration limited to 300 a year

7,000

Under German occupation or control by December 1941. Home of 8 million Jews

Countries almost entirely closed to Jewish refugees

Countries restricting Jewish immigration by means of a quota system or other barriers

Countries partially, but not entirely restricting Jewish immigration

⊕ City accepting more Jewish refugees than were allowed to enter Canada, Australia, New Zealand, British Africa, South Africa and India (all in the British Empire) combined

5,000 Number of Jewish refugees allowed to enter 1933-1945. Shanghai alone required no visa

14,000 **50,000**

January 1939 Anglo-American suggestion that Jewish refugees go to Angola not followed up for fear of offending the Portuguese

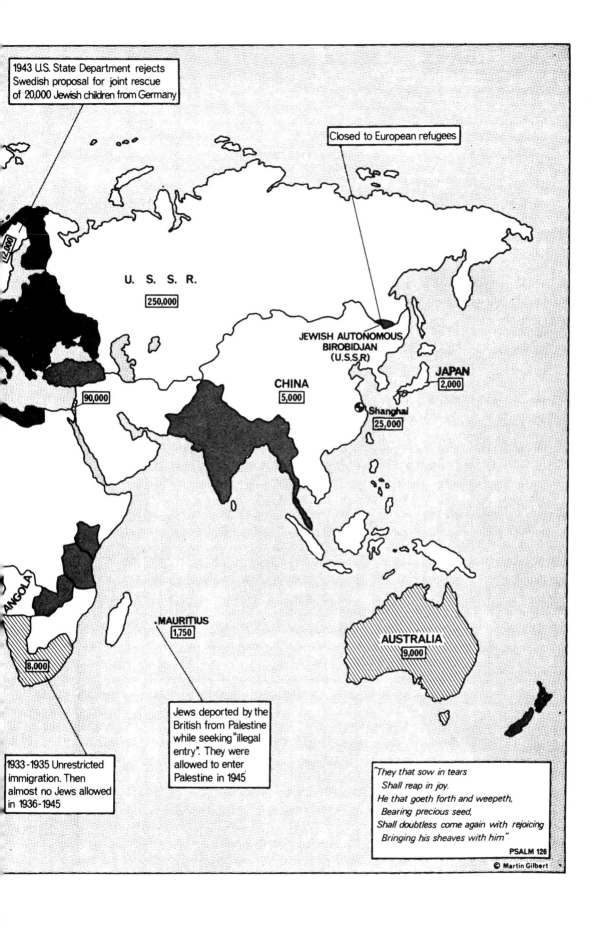

1943 U.S. State Department rejects Swedish proposal for joint rescue of 20,000 Jewish children from Germany

Closed to European refugees

12,000

U. S. S. R.
250,000

JEWISH AUTONOMOUS BIROBIDJAN (U.S.S.R)

JAPAN
2,000

CHINA
5,000

Shanghai
25,000

90,000

ANGOLA

MAURITIUS
1,750

AUSTRALIA
9,000

8,000

1933-1935 Unrestricted immigration. Then almost no Jews allowed in 1936-1945

Jews deported by the British from Palestine while seeking "illegal entry". They were allowed to enter Palestine in 1945

"They that sow in tears
    Shall reap in joy.
He that goeth forth and weepeth,
    Bearing precious seed,
Shall doubtless come again with rejoicing
    Bringing his sheaves with him"

PSALM 126

© Martin Gilbert

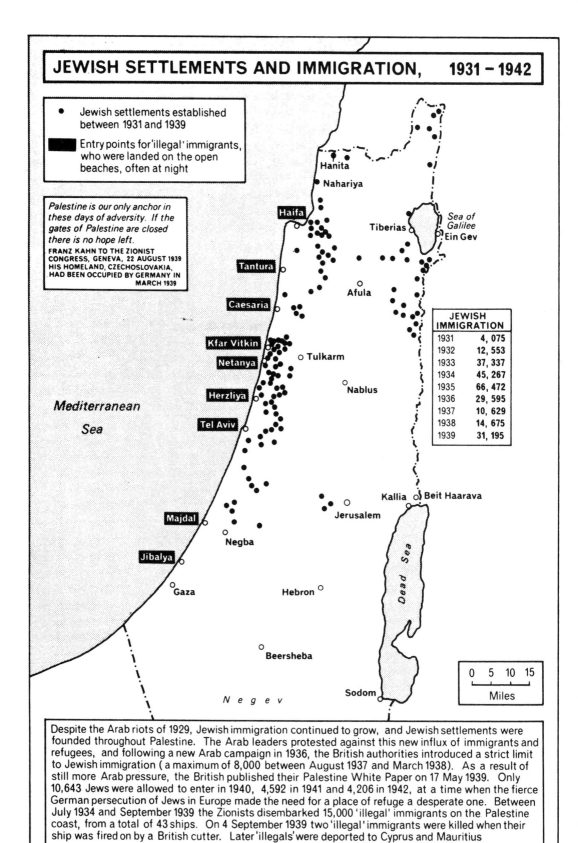

JEWISH SETTLEMENTS AND IMMIGRATION, 1931 – 1942

- Jewish settlements established between 1931 and 1939
- Entry points for 'illegal' immigrants, who were landed on the open beaches, often at night

Palestine is our only anchor in these days of adversity. If the gates of Palestine are closed there is no hope left.
FRANZ KAHN TO THE ZIONIST CONGRESS, GENEVA, 22 AUGUST 1939 HIS HOMELAND, CZECHOSLOVAKIA, HAD BEEN OCCUPIED BY GERMANY IN MARCH 1939

Mediterranean Sea

Hanita
Nahariya
Haifa
Tiberias
Sea of Galilee
Ein Gev
Tantura
Caesaria
Afula
Kfar Vitkin
Netanya
Tulkarm
Herzliya
Nablus
Tel Aviv
Kallia
Beit Haarava
Jerusalem
Majdal
Dead Sea
Negba
Jibalya
Gaza
Hebron
Beersheba
Sodom
N e g e v

JEWISH IMMIGRATION	
1931	4, 075
1932	12, 553
1933	37, 337
1934	45, 267
1935	66, 472
1936	29, 595
1937	10, 629
1938	14, 675
1939	31, 195

0 5 10 15
Miles

Despite the Arab riots of 1929, Jewish immigration continued to grow, and Jewish settlements were founded throughout Palestine. The Arab leaders protested against this new influx of immigrants and refugees, and following a new Arab campaign in 1936, the British authorities introduced a strict limit to Jewish immigration (a maximum of 8,000 between August 1937 and March 1938). As a result of still more Arab pressure, the British published their Palestine White Paper on 17 May 1939. Only 10,643 Jews were allowed to enter in 1940, 4,592 in 1941 and 4,206 in 1942, at a time when the fierce German persecution of Jews in Europe made the need for a place of refuge a desperate one. Between July 1934 and September 1939 the Zionists disembarked 15,000 'illegal' immigrants on the Palestine coast, from a total of 43 ships. On 4 September 1939 two 'illegal' immigrants were killed when their ship was fired on by a British cutter. Later 'illegals' were deported to Cyprus and Mauritius

© Martin Gilbert

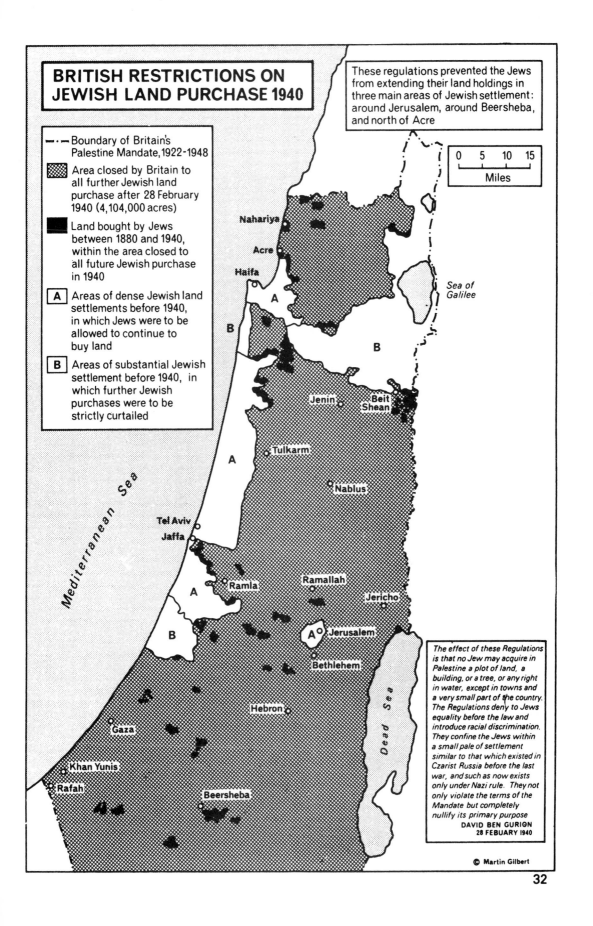

BRITISH RESTRICTIONS ON JEWISH LAND PURCHASE 1940

These regulations prevented the Jews from extending their land holdings in three main areas of Jewish settlement: around Jerusalem, around Beersheba, and north of Acre

–··– Boundary of Britain's Palestine Mandate, 1922-1948

Area closed by Britain to all further Jewish land purchase after 28 February 1940 (4,104,000 acres)

Land bought by Jews between 1880 and 1940, within the area closed to all future Jewish purchase in 1940

A Areas of dense Jewish land settlements before 1940, in which Jews were to be allowed to continue to buy land

B Areas of substantial Jewish settlement before 1940, in which further Jewish purchases were to be strictly curtailed

0 5 10 15
Miles

Nahariya

Acre

Haifa

A

B

Sea of Galilee

B

Jenin   Beit Shean

Tulkarm

Nablus

A

Tel Aviv

Jaffa

Mediterranean Sea

Ramla   **A**

Ramallah

Jericho

B

A Jerusalem

Bethlehem

Hebron

Dead Sea

Gaza

Khan Yunis

Rafah

Beersheba

The effect of these Regulations is that no Jew may acquire in Palestine a plot of land, a building, or a tree, or any right in water, except in towns and a very small part of the country. The Regulations deny to Jews equality before the law and introduce racial discrimination. They confine the Jews within a small pale of settlement similar to that which existed in Czarist Russia before the last war, and such as now exists only under Nazi rule. They not only violate the terms of the Mandate but completely nullify its primary purpose
DAVID BEN GURION
28 FEBUARY 1940

© Martin Gilbert

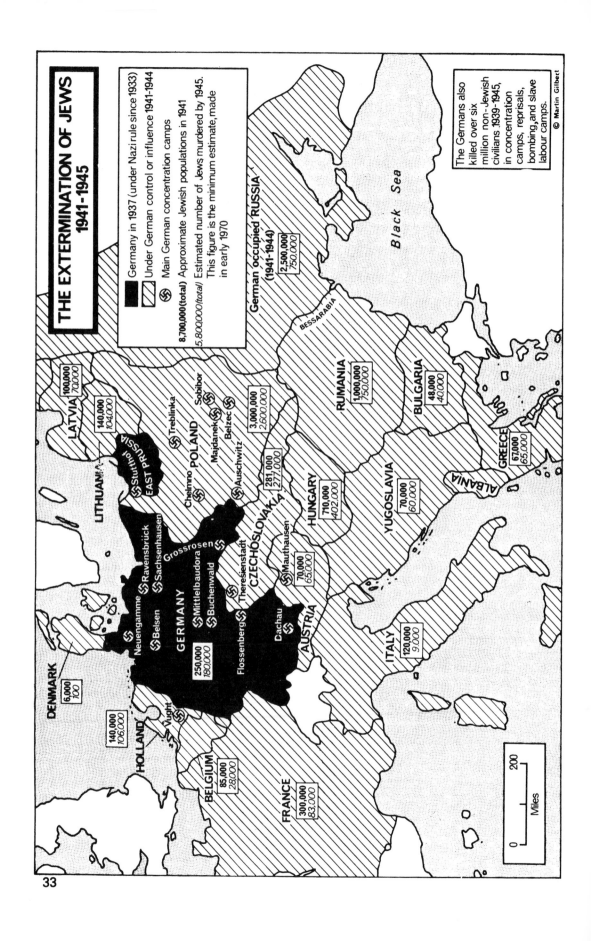

THE EXTERMINATION OF JEWS
1941-1945

Germany in 1937 (under Nazi rule since 1933)

Under German control or influence 1941-1944

卐 Main German concentration camps

8,700,000 (total) Approximate Jewish populations in 1941

5,800,000 (total) Estimated number of Jews murdered by 1945. This figure is the minimum estimate, made in early 1970

The Germans also killed over six million non-Jewish civilians ,1939-1945, in concentration camps, reprisals, bombing, and slave labour camps.

© Martin Gilbert

German occupied RUSSIA (1941-1944)

2,500,000
750,000

Black Sea

BESSARABIA

DENMARK
6,000
100

HOLLAND
140,000
106,000

Vught 卐

BELGIUM
85,000
28,000

FRANCE
300,000
83,000

卐 Neuengamme
卐 Belsen
卐 Ravensbrück
卐 Sachsenhausen
卐 Grossrosen

GERMANY
250,000
180,000

卐 Mittleibaudora
卐 Buchenwald
卐 Flossenberg
Dachau 卐

AUSTRIA

ITALY
120,000
9,000

Stutthof 卐
EAST PRUSSIA

LITHUANIA

LATVIA
100,000
70,000

140,000
104,000

卐 Treblinka

POLAND

Chelmno 卐
Sobibor 卐
Majdanek 卐
Belzec 卐
Auschwitz 卐

3,000,000
2,600,000

Theresienstadt 卐
CZECHOSLOVAKIA
281,000
277,000

卐 Mauthausen
70,000
65,000

HUNGARY
710,000
402,000

RUMANIA
1,000,000
750,000

YUGOSLAVIA
70,000
60,000

BULGARIA
48,000
40,000

ALBANIA

GREECE
67,000
65,000

0   200
Miles

33

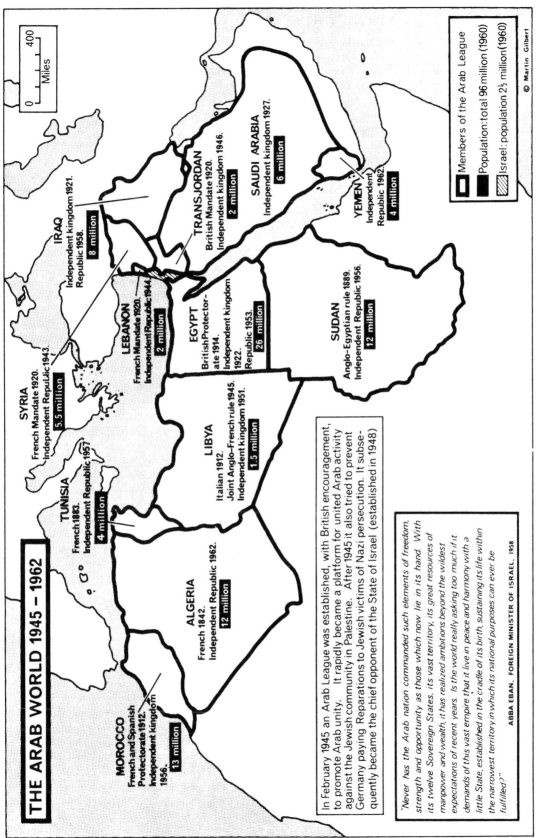

THE ARAB WORLD 1945 – 1962

400
0
Miles

MOROCCO
French and Spanish Protectorate 1912.
Independent kingdom 1956.
13 million

ALGERIA
French 1842.
Independent Republic 1962.
12 million

TUNISIA
French 1883.
Independent Republic 1957
4 million

SYRIA
French Mandate 1920.
Independent Republic 1943.
5.5 million

IRAQ
Independent kingdom 1921.
Republic 1958.
8 million

LEBANON
French Mandate 1920.
Independent Republic 1944.
2 million

LIBYA
Italian 1912.
Joint Anglo–French rule 1945.
Independent kingdom 1951.
1.5 million

EGYPT
British Protectorate 1914.
Independent kingdom 1922.
Republic 1953.
26 million

TRANSJORDAN
British Mandate 1920.
Independent kingdom 1946.
2 million

SAUDI ARABIA
Independent kingdom 1927.
6 million

YEMEN
Independent Republic 1962.
4 million

SUDAN
Anglo–Egyptian rule 1889.
Independent Republic 1956.
12 million

◻ Members of the Arab League

◼ Population: total 96 million (1960)

▨ Israel: population 2½ million (1960)

© Martin Gilbert

In February 1945 an Arab League was established, with British encouragement, to promote Arab unity. It rapidly became a platform for united Arab activity against the Jewish community in Palestine. After 1945 it also tried to prevent Germany paying Reparations to Jewish victims of Nazi persecution. It subsequently became the chief opponent of the State of Israel (established in 1948)

"Never has the Arab nation commanded such elements of freedom, strength and opportunity as those which now lie in its hand. With its twelve Sovereign States, its vast territory, its great resources of manpower and wealth, it has realized ambitions beyond the wildest expectations of recent years. Is the world really asking too much if it demands of this vast empire that it live in peace and harmony with a little State, established in the cradle of its birth, sustaining its life within the narrowest territory in which its national purposes can ever be fulfilled?"

ABBA EBAN, FOREIGN MINISTER OF ISRAEL, 1958

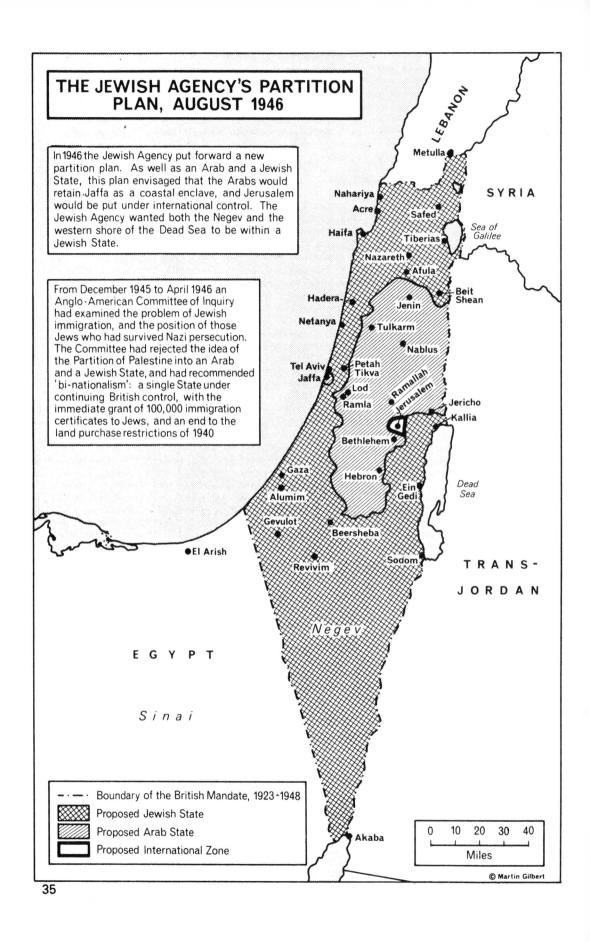

THE JEWISH AGENCY'S PARTITION PLAN, AUGUST 1946

In 1946 the Jewish Agency put forward a new partition plan. As well as an Arab and a Jewish State, this plan envisaged that the Arabs would retain Jaffa as a coastal enclave, and Jerusalem would be put under international control. The Jewish Agency wanted both the Negev and the western shore of the Dead Sea to be within a Jewish State.

From December 1945 to April 1946 an Anglo-American Committee of Inquiry had examined the problem of Jewish immigration, and the position of those Jews who had survived Nazi persecution. The Committee had rejected the idea of the Partition of Palestine into an Arab and a Jewish State, and had recommended 'bi-nationalism': a single State under continuing British control, with the immediate grant of 100,000 immigration certificates to Jews, and an end to the land purchase restrictions of 1940

LEBANON

SYRIA

Metulla

Nahariya
Acre
Safed
Haifa
Sea of Galilee
Tiberias
Nazareth
Afula
Beit Shean
Hadera
Jenin
Netanya
Tulkarm
Nablus
Tel Aviv
Petah Tikva
Jaffa
Ramallah
Lod
Jerusalem
Jericho
Ramla
Kallia
Bethlehem
Gaza
Hebron
Alumim
Ein Gedi
Dead Sea
Gevulot
Beersheba
El Arish
Sodom
Revivim

TRANS-
JORDAN

Negev

EGYPT

Sinai

Akaba

Boundary of the British Mandate, 1923-1948
Proposed Jewish State
Proposed Arab State
Proposed International Zone

0   10   20   30   40
Miles

© Martin Gilbert

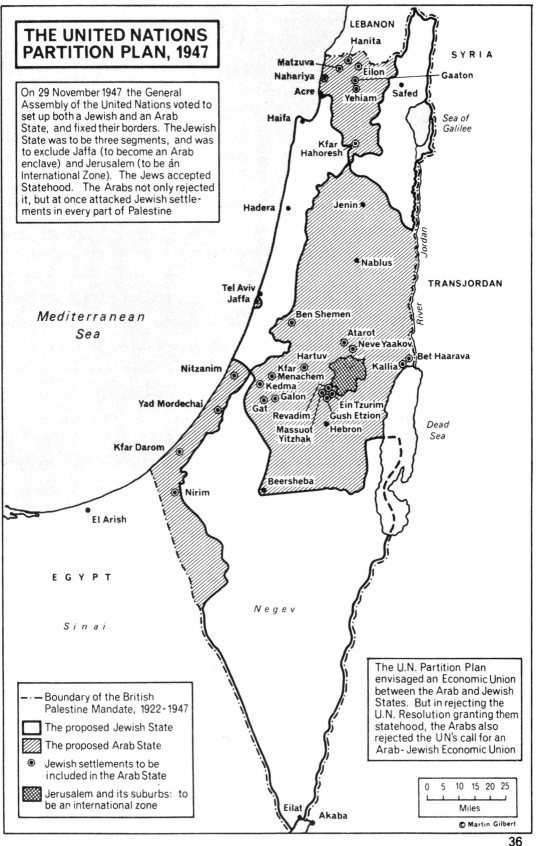

THE UNITED NATIONS PARTITION PLAN, 1947

On 29 November 1947 the General Assembly of the United Nations voted to set up both a Jewish and an Arab State, and fixed their borders. The Jewish State was to be three segments, and was to exclude Jaffa (to become an Arab enclave) and Jerusalem (to be an International Zone). The Jews accepted Statehood. The Arabs not only rejected it, but at once attacked Jewish settlements in every part of Palestine

LEBANON

Hanita

Matzuva
Nahariya
Acre

Eilon

Gaaton

SYRIA

Yehiam

Safed

Sea of Galilee

Haifa

Kfar Hahoresh

Jenin

Hadera

Nablus

Jordan River

TRANSJORDAN

Tel Aviv
Jaffa

Ben Shemen

Atarot
Neve Yaakov

Mediterranean Sea

Hartuv
Kfar Menachem
Kedma
Galon
Gat
Revadim

Kallia

Bet Haarava

Nitzanim

Ein Tzurim
Gush Etzion
Hebron

Dead Sea

Yad Mordechai

Massuot Yitzhak

Kfar Darom

Beersheba

Nirim

El Arish

EGYPT

Negev

Sinai

The U.N. Partition Plan envisaged an Economic Union between the Arab and Jewish States. But in rejecting the U.N. Resolution granting them statehood, the Arabs also rejected the UN's call for an Arab-Jewish Economic Union

Boundary of the British Palestine Mandate, 1922-1947

The proposed Jewish State

The proposed Arab State

⊙ Jewish settlements to be included in the Arab State

Jerusalem and its suburbs: to be an international zone

0 5 10 15 20 25
Miles

Eilat
Akaba

© Martin Gilbert

THE IMMEDIATE RESPONSE TO THE UNITED NATIONS PARTITION PLAN 30 NOVEMBER – 11 DECEMBER 1947

0   10   20
Miles

2 Dec 1 Jew killed by Arab rioters
7 Dec 1 Jew killed by Arabs. 1 Arab killed after Arabs attack Jews
8 Dec 5 Jews, 1 Arab and 1 British policeman killed
9 Dec 2 Jews, 2 Arabs killed after an Arab boy threw a grenade at a Jewish shop
11 Dec Jews attack Arab quarter, 6 Arabs and 1 Jew killed

On 1 December 1947 an Arab mob in Beirut (Lebanon) attacked Jewish houses and synagogues, which were looted and burned. On 9 December an Arab mob in Aden attacked the Jewish community there, killing 82 Jews. During the fighting, 34 Arabs were killed

30 Nov 7 Jews murdered by Arabs who attacked Jerusalem-bound buses
2 Dec 3 Jews killed
7 Dec 4 Jews and 1 Arab killed
8 Dec 6 Jews murdered
11 Dec 1 Jew killed when Arabs attack Jewish convoy

4 Dec 1 Jew killed

4 Dec 1 Arab killed

6 Dec Arabs kill 8 Jews
7 Dec Arabs kill 5 Jews
8 Dec 4 Jews killed or burnt alive, during an Arab attack

2 Dec 2 Arabs killed
9 Dec 2 Jewish guards killed

6 Dec Arabs attack a Jewish settlement; 8 Arabs killed

2 Dec 1 Jew killed by Arabs

9 Dec A Jewish watchman killed by Arabs

7 Dec Arabs ambush a Jewish bus. A Jewish girl of 19 killed

Haifa

Hadera

Kfar Saba

Tel Aviv
Jaffa
Salameh
Efal
Ramla
Rehovot
Abu Ghosh
Jerusalem

Hebron

Dead Sea

Gevulot

9 Dec 6 teenage Jews (one a girl) murdered by an Arab mob

11 Dec Arabs attack a Jewish convoy on the Jerusalem–Hebron road. 10 Jews killed

2 Dec A mob of 200 Arabs loot Jewish shops, smash windows of houses and stab passers-by. 6 Jews seriously injured
3 Dec 1 Jew killed by Arabs in the Old City
4 Dec Arabs attack a Jewish synagogue: 1 Jew wounded, 1 Arab killed
7 Dec 2 Jews killed by Arab snipers
8 Dec 2 Jews stabbed to death
11 Dec Arabs attempt to drive the Jews from the Old City. The Jews defend themselves. 3 Arabs and 1 Jew killed

In the twelve days following the United Nations Partition plan, 79 Jews were killed by Arabs throughout Palestine. The British, who were still responsible for law and order, did not always enforce it. Sometimes they would disarm a Jewish defence group, which would then be attacked by armed Arabs. The Jews defended themselves, and in places counter-attacked. During this same period, 32 Arabs were killed, some by Jews and some by the British police

Arabs and Moslems throughout the world will obstruct it, and all Asia with its thousand million people will oppose it **THE SYRIAN DELEGATE TO THE UN, ON THE UN PARTITION PLAN 30 NOV 1947**

Jews will take all measures to protect themselves **THE JEWISH NATIONAL COUNCIL, 3 DEC 1947**

© Martin Gilbert

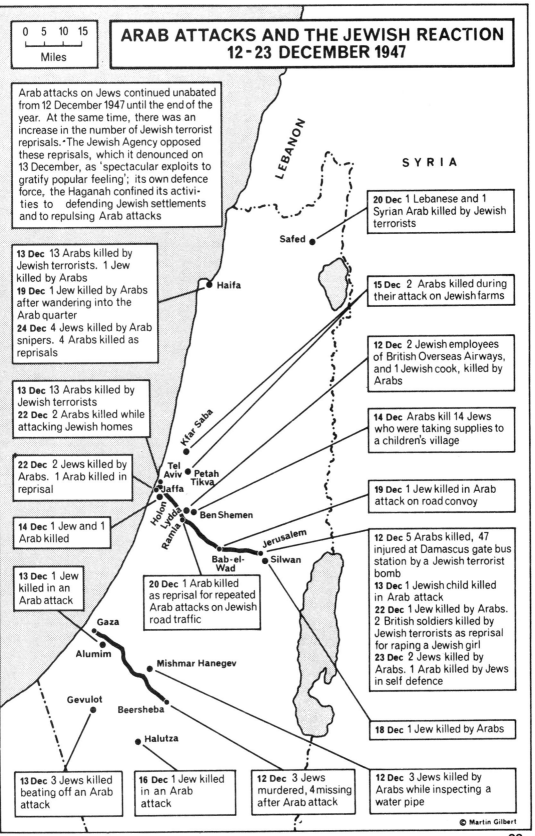

ARAB ATTACKS AND THE JEWISH REACTION 12-23 DECEMBER 1947

0 5 10 15
Miles

LEBANON

SYRIA

Arab attacks on Jews continued unabated from 12 December 1947 until the end of the year. At the same time, there was an increase in the number of Jewish terrorist reprisals. The Jewish Agency opposed these reprisals, which it denounced on 13 December, as 'spectacular exploits to gratify popular feeling'; its own defence force, the Haganah confined its activities to defending Jewish settlements and to repulsing Arab attacks

20 Dec 1 Lebanese and 1 Syrian Arab killed by Jewish terrorists

Safed

13 Dec 13 Arabs killed by Jewish terrorists. 1 Jew killed by Arabs
19 Dec 1 Jew killed by Arabs after wandering into the Arab quarter
24 Dec 4 Jews killed by Arab snipers. 4 Arabs killed as reprisals

Haifa

15 Dec 2 Arabs killed during their attack on Jewish farms

12 Dec 2 Jewish employees of British Overseas Airways, and 1 Jewish cook, killed by Arabs

13 Dec 13 Arabs killed by Jewish terrorists
22 Dec 2 Arabs killed while attacking Jewish homes

14 Dec Arabs kill 14 Jews who were taking supplies to a children's village

Kfar Saba

22 Dec 2 Jews killed by Arabs. 1 Arab killed in reprisal

Tel Aviv • Petah Tikva
• Jaffa

19 Dec 1 Jew killed in Arab attack on road convoy

14 Dec 1 Jew and 1 Arab killed

Holon
Lydda
Ramla
Ben Shemen

Jerusalem

12 Dec 5 Arabs killed, 47 injured at Damascus gate bus station by a Jewish terrorist bomb
13 Dec 1 Jewish child killed in Arab attack
22 Dec 1 Jew killed by Arabs. 2 British soldiers killed by Jewish terrorists as reprisal for raping a Jewish girl
23 Dec 2 Jews killed by Arabs. 1 Arab killed by Jews in self defence

13 Dec 1 Jew killed in an Arab attack

Gaza

20 Dec 1 Arab killed as reprisal for repeated Arab attacks on Jewish road traffic

Bab-el-Wad
• Silwan

Alumim

• Mishmar Hanegev

Gevulot

18 Dec 1 Jew killed by Arabs

Beersheba

• Halutza

13 Dec 3 Jews killed beating off an Arab attack

16 Dec 1 Jew killed in an Arab attack

12 Dec 3 Jews murdered, 4 missing after Arab attack

12 Dec 3 Jews killed by Arabs while inspecting a water pipe

© Martin Gilbert

38

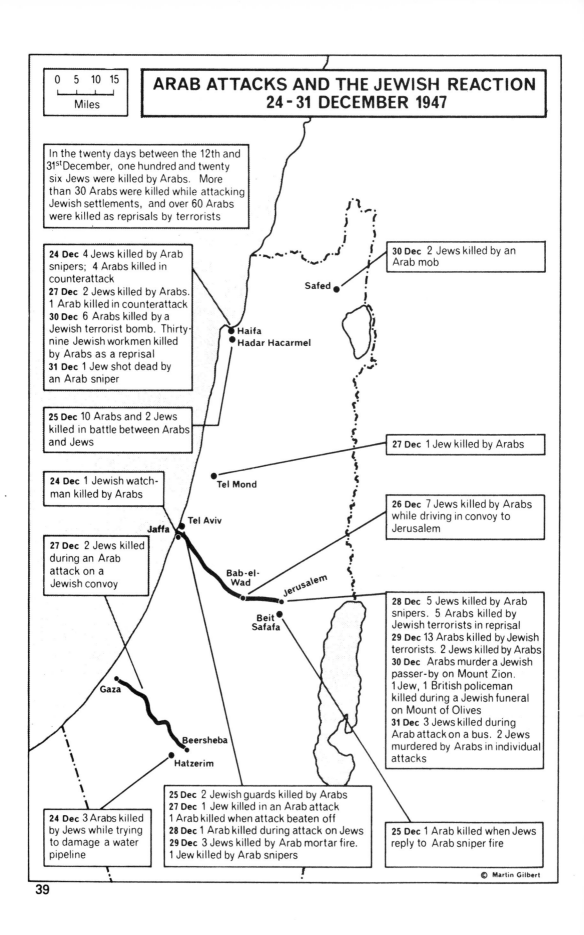

ARAB ATTACKS AND THE JEWISH REACTION
24 - 31 DECEMBER 1947

0 5 10 15
Miles

In the twenty days between the 12th and 31st December, one hundred and twenty six Jews were killed by Arabs. More than 30 Arabs were killed while attacking Jewish settlements, and over 60 Arabs were killed as reprisals by terrorists

24 Dec 4 Jews killed by Arab snipers; 4 Arabs killed in counterattack
27 Dec 2 Jews killed by Arabs. 1 Arab killed in counterattack
30 Dec 6 Arabs killed by a Jewish terrorist bomb. Thirty-nine Jewish workmen killed by Arabs as a reprisal
31 Dec 1 Jew shot dead by an Arab sniper

25 Dec 10 Arabs and 2 Jews killed in battle between Arabs and Jews

24 Dec 1 Jewish watchman killed by Arabs

27 Dec 2 Jews killed during an Arab attack on a Jewish convoy

30 Dec 2 Jews killed by an Arab mob

Safed

27 Dec 1 Jew killed by Arabs

Haifa
Hadar Hacarmel

Tel Mond

26 Dec 7 Jews killed by Arabs while driving in convoy to Jerusalem

Tel Aviv
Jaffa

Bab-el-Wad

Jerusalem

Beit Safafa

28 Dec 5 Jews killed by Arab snipers. 5 Arabs killed by Jewish terrorists in reprisal
29 Dec 13 Arabs killed by Jewish terrorists. 2 Jews killed by Arabs
30 Dec Arabs murder a Jewish passer-by on Mount Zion. 1 Jew, 1 British policeman killed during a Jewish funeral on Mount of Olives
31 Dec 3 Jews killed during Arab attack on a bus. 2 Jews murdered by Arabs in individual attacks

Gaza

Beersheba
Hatzerim

24 Dec 3 Arabs killed by Jews while trying to damage a water pipeline

25 Dec 2 Jewish guards killed by Arabs
27 Dec 1 Jew killed in an Arab attack
1 Arab killed when attack beaten off
28 Dec 1 Arab killed during attack on Jews
29 Dec 3 Jews killed by Arab mortar fire.
1 Jew killed by Arab snipers

25 Dec 1 Arab killed when Jews reply to Arab sniper fire

© Martin Gilbert

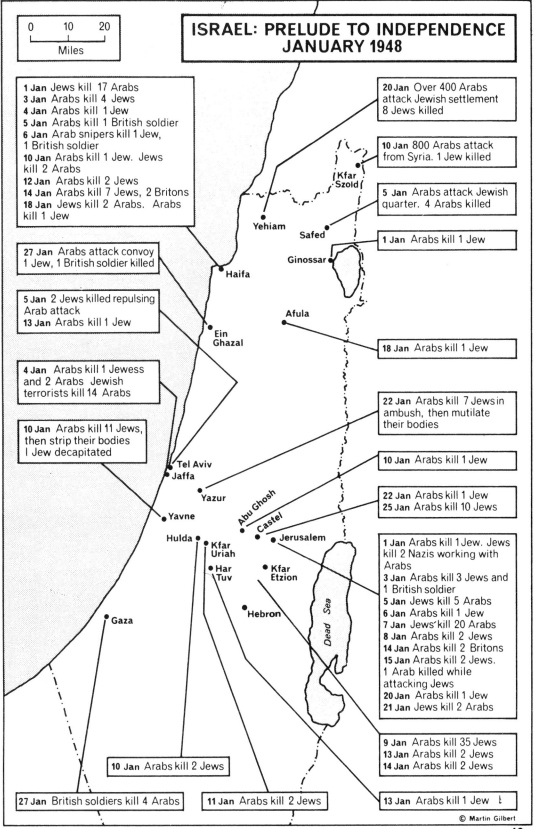

ISRAEL: PRELUDE TO INDEPENDENCE JANUARY 1948

Miles
0   10   20

1 Jan Jews kill 17 Arabs
3 Jan Arabs kill 4 Jews
4 Jan Arabs kill 1 Jew
5 Jan Arabs kill 1 British soldier
6 Jan Arab snipers kill 1 Jew, 1 British soldier
10 Jan Arabs kill 1 Jew. Jews kill 2 Arabs
12 Jan Arabs kill 2 Jews
14 Jan Arabs kill 7 Jews, 2 Britons
18 Jan Jews kill 2 Arabs. Arabs kill 1 Jew

27 Jan Arabs attack convoy 1 Jew, 1 British soldier killed

5 Jan 2 Jews killed repulsing Arab attack
13 Jan Arabs kill 1 Jew

4 Jan Arabs kill 1 Jewess and 2 Arabs Jewish terrorists kill 14 Arabs

10 Jan Arabs kill 11 Jews, then strip their bodies I Jew decapitated

20 Jan Over 400 Arabs attack Jewish settlement 8 Jews killed

10 Jan 800 Arabs attack from Syria. 1 Jew killed

5 Jan Arabs attack Jewish quarter. 4 Arabs killed

1 Jan Arabs kill 1 Jew

18 Jan Arabs kill 1 Jew

22 Jan Arabs kill 7 Jews in ambush, then mutilate their bodies

10 Jan Arabs kill 1 Jew

22 Jan Arabs kill 1 Jew
25 Jan Arabs kill 10 Jews

1 Jan Arabs kill 1 Jew. Jews kill 2 Nazis working with Arabs
3 Jan Arabs kill 3 Jews and 1 British soldier
5 Jan Jews kill 5 Arabs
6 Jan Arabs kill 1 Jew
7 Jan Jews kill 20 Arabs
8 Jan Arabs kill 2 Jews
14 Jan Arabs kill 2 Britons
15 Jan Arabs kill 2 Jews. 1 Arab killed while attacking Jews
20 Jan Arabs kill 1 Jew
21 Jan Jews kill 2 Arabs

9 Jan Arabs kill 35 Jews
13 Jan Arabs kill 2 Jews
14 Jan Arabs kill 2 Jews

10 Jan Arabs kill 2 Jews

27 Jan British soldiers kill 4 Arabs

11 Jan Arabs kill 2 Jews

13 Jan Arabs kill 1 Jew

Kfar Szold
Yehiam
Safed
Ginossar
Haifa
Afula
Ein Ghazal
Tel Aviv
Jaffa
Yazur
Abu Ghosh
Castel
Yavne
Hulda
Kfar Uriah
Har Tuv
Kfar Etzion
Jerusalem
Gaza
Hebron
Dead Sea

© Martin Gilbert

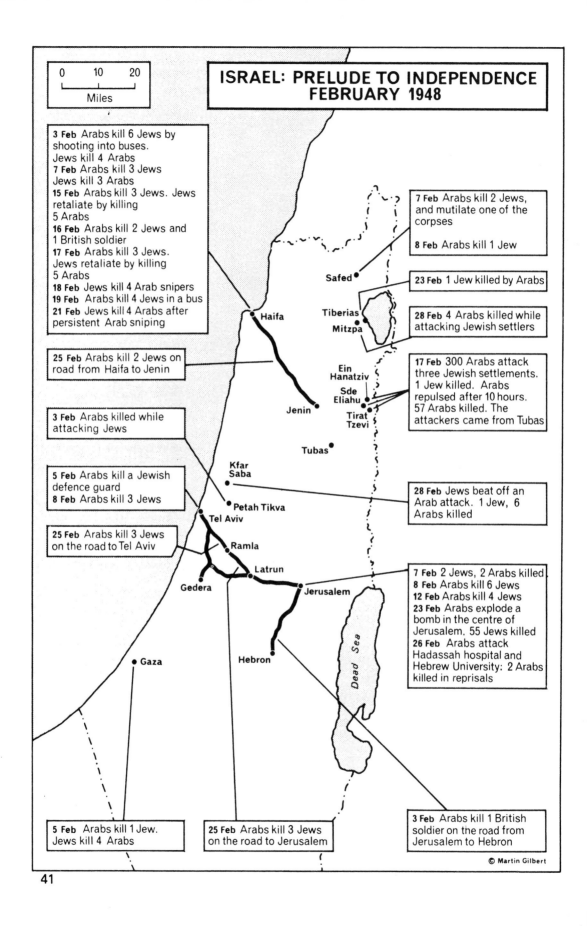

ISRAEL: PRELUDE TO INDEPENDENCE FEBRUARY 1948

0 10 20
Miles

3 Feb Arabs kill 6 Jews by shooting into buses.
Jews kill 4 Arabs
7 Feb Arabs kill 3 Jews
Jews kill 3 Arabs
15 Feb Arabs kill 3 Jews. Jews retaliate by killing
5 Arabs
16 Feb Arabs kill 2 Jews and 1 British soldier
17 Feb Arabs kill 3 Jews. Jews retaliate by killing
5 Arabs
18 Feb Jews kill 4 Arab snipers
19 Feb Arabs kill 4 Jews in a bus
21 Feb Jews kill 4 Arabs after persistent Arab sniping

25 Feb Arabs kill 2 Jews on road from Haifa to Jenin

3 Feb Arabs killed while attacking Jews

5 Feb Arabs kill a Jewish defence guard
8 Feb Arabs kill 3 Jews

25 Feb Arabs kill 3 Jews on the road to Tel Aviv

7 Feb Arabs kill 2 Jews, and mutilate one of the corpses
8 Feb Arabs kill 1 Jew

23 Feb 1 Jew killed by Arabs

28 Feb 4 Arabs killed while attacking Jewish settlers

17 Feb 300 Arabs attack three Jewish settlements. 1 Jew killed. Arabs repulsed after 10 hours. 57 Arabs killed. The attackers came from Tubas

28 Feb Jews beat off an Arab attack. 1 Jew, 6 Arabs killed

7 Feb 2 Jews, 2 Arabs killed
8 Feb Arabs kill 6 Jews
12 Feb Arabs kill 4 Jews
23 Feb Arabs explode a bomb in the centre of Jerusalem. 55 Jews killed
26 Feb Arabs attack Hadassah hospital and Hebrew University: 2 Arabs killed in reprisals

5 Feb Arabs kill 1 Jew. Jews kill 4 Arabs

25 Feb Arabs kill 3 Jews on the road to Jerusalem

3 Feb Arabs kill 1 British soldier on the road from Jerusalem to Hebron

Safed

Tiberias
Mitzpa

Haifa

Ein Hanatziv
Sde Eliahu
Tirat Tzevi

Jenin

Tubas

Kfar Saba

Petah Tikva
Tel Aviv

Ramla

Latrun

Gedera

Jerusalem

Hebron

Gaza

Dead Sea

© Martin Gilbert

41

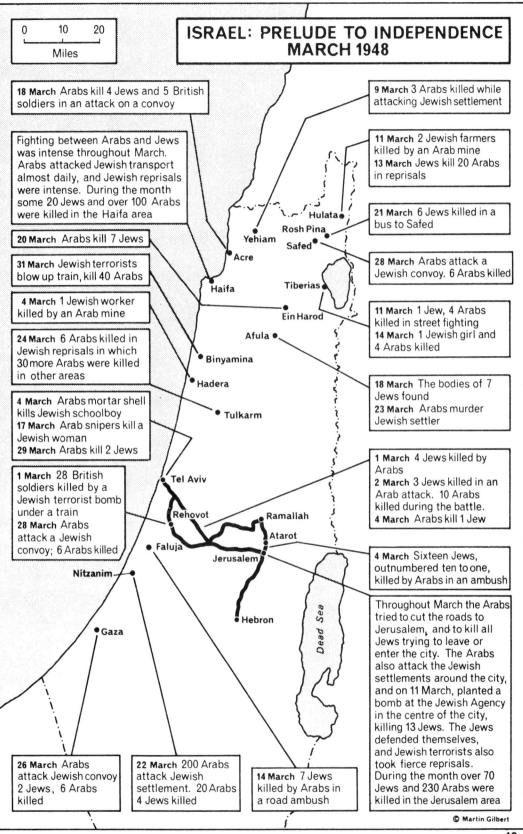

ISRAEL: PRELUDE TO INDEPENDENCE MARCH 1948

0 10 20
Miles

18 March Arabs kill 4 Jews and 5 British soldiers in an attack on a convoy

Fighting between Arabs and Jews was intense throughout March. Arabs attacked Jewish transport almost daily, and Jewish reprisals were intense. During the month some 20 Jews and over 100 Arabs were killed in the Haifa area

20 March Arabs kill 7 Jews

31 March Jewish terrorists blow up train, kill 40 Arabs

4 March 1 Jewish worker killed by an Arab mine

24 March 6 Arabs killed in Jewish reprisals in which 30 more Arabs were killed in other areas

4 March Arabs mortar shell kills Jewish schoolboy
17 March Arab snipers kill a Jewish woman
29 March Arabs kill 2 Jews

1 March 28 British soldiers killed by a Jewish terrorist bomb under a train
28 March Arabs attack a Jewish convoy; 6 Arabs killed

9 March 3 Arabs killed while attacking Jewish settlement

11 March 2 Jewish farmers killed by an Arab mine
13 March Jews kill 20 Arabs in reprisals

21 March 6 Jews killed in a bus to Safed

28 March Arabs attack a Jewish convoy. 6 Arabs killed

11 March 1 Jew, 4 Arabs killed in street fighting
14 March 1 Jewish girl and 4 Arabs killed

18 March The bodies of 7 Jews found
23 March Arabs murder Jewish settler

1 March 4 Jews killed by Arabs
2 March 3 Jews killed in an Arab attack. 10 Arabs killed during the battle.
4 March Arabs kill 1 Jew

4 March Sixteen Jews, outnumbered ten to one, killed by Arabs in an ambush

Throughout March the Arabs tried to cut the roads to Jerusalem, and to kill all Jews trying to leave or enter the city. The Arabs also attack the Jewish settlements around the city, and on 11 March, planted a bomb at the Jewish Agency in the centre of the city, killing 13 Jews. The Jews defended themselves, and Jewish terrorists also took fierce reprisals. During the month over 70 Jews and 230 Arabs were killed in the Jerusalem area

26 March Arabs attack Jewish convoy 2 Jews, 6 Arabs killed

22 March 200 Arabs attack Jewish settlement. 20 Arabs 4 Jews killed

14 March 7 Jews killed by Arabs in a road ambush

Hulata
Rosh Pina
Yehiam
Safed
Acre
Tiberias
Haifa
Ein Harod
Afula
Binyamina
Hadera
Tulkarm
Tel Aviv
Rehovot
Ramallah
Atarot
Faluja
Jerusalem
Nitzanim
Hebron
Gaza
Dead Sea

© Martin Gilbert

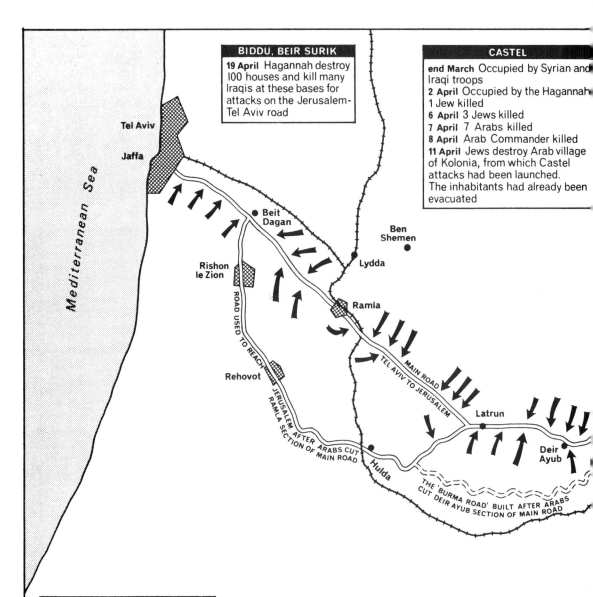

BIDDU, BEIR SURIK

19 April Hagannah destroy 100 houses and kill many Iraqis at these bases for attacks on the Jerusalem-Tel Aviv road

CASTEL

end March Occupied by Syrian and Iraqi troops
2 April Occupied by the Hagannah 1 Jew killed
6 April 3 Jews killed
7 April 7 Arabs killed
8 April Arab Commander killed
11 April Jews destroy Arab village of Kolonia, from which Castel attacks had been launched. The inhabitants had already been evacuated

Tel Aviv

Jaffa

Mediterranean Sea

Beit Dagan

Ben Shemen

Lydda

Rishon le Zion

Ramla

ROAD USED TO REACH JERUSALEM RAMLA SECTION OF MAIN ROAD

Rehovot

JERUSALEM AFTER ARABS CUT SECTION OF MAIN ROAD

MAIN ROAD

TEL AVIV TO JERUSALEM

Latrun

Deir Ayub

Hulda

THE 'BURMA ROAD' BUILT AFTER ARABS CUT DEIR AYUB SECTION OF MAIN ROAD

RAMLA

12 April Hagannah blow up 12 buildings in area from which Jerusalem road attacks had come

In the six weeks before the British withdrew from Palestine, the Arabs did everything in their power to prevent the Jews from reaching Jerusalem, and sought to disrupt all Jewish life within the city. Many of the Arabs involved were regular soldiers from Syria and Iraq

 Constant Arab sniping throughout April and May 1948 against vehicles on the roads to Jerusalem

© Martin Gilbert

LYDDA

5 April Hagannah kill 10 Iraqi soldiers in camp from which the Jerusalam road had been under attack

DEIR AYUB

20 April 6 Jews killed in an Arab ambush

DEIR YASSIN

9 April Jewish terrorists massacre over 200 Arabs. The Jewish Agency and the Hagannah both immediately condemned the killings as 'utterly repugnant'

SARIS

16 April Jews capture Syrian army base. Several dozen Syrians, and 3 Jews killed

KFAR ETZION

13 April Attack by 400 troops repulsed
20-30 April Jews repel repeated Arab attacks
4 May Arab attacks beaten off; 12 Arabs killed
12 May Several hundred Arabs renew the attack. 100 Jews killed. Only 4 survive. 15 Jews were machine gunned to death after they had surrendered, and were being photographed by their captors

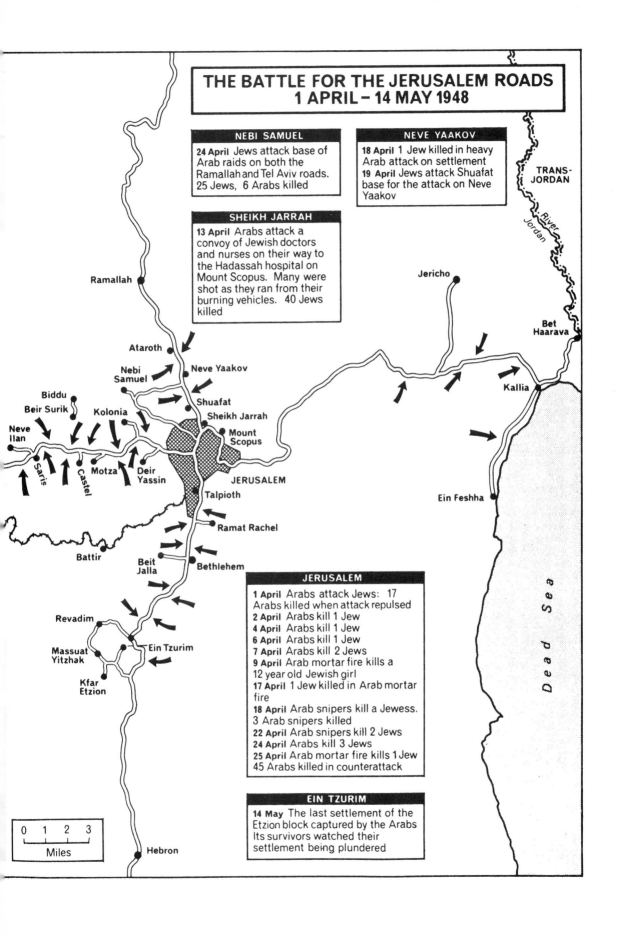

THE BATTLE FOR THE JERUSALEM ROADS
1 APRIL – 14 MAY 1948

NEBI SAMUEL
24 April Jews attack base of Arab raids on both the Ramallah and Tel Aviv roads. 25 Jews, 6 Arabs killed

NEVE YAAKOV
18 April 1 Jew killed in heavy Arab attack on settlement
19 April Jews attack Shuafat base for the attack on Neve Yaakov

SHEIKH JARRAH
13 April Arabs attack a convoy of Jewish doctors and nurses on their way to the Hadassah hospital on Mount Scopus. Many were shot as they ran from their burning vehicles. 40 Jews killed

JERUSALEM
1 April Arabs attack Jews: 17 Arabs killed when attack repulsed
2 April Arabs kill 1 Jew
4 April Arabs kill 1 Jew
6 April Arabs kill 1 Jew
7 April Arabs kill 2 Jews
9 April Arab mortar fire kills a 12 year old Jewish girl
17 April 1 Jew killed in Arab mortar fire
18 April Arab snipers kill a Jewess. 3 Arab snipers killed
22 April Arab snipers kill 2 Jews
24 April Arabs kill 3 Jews
25 April Arab mortar fire kills 1 Jew 45 Arabs killed in counterattack

EIN TZURIM
14 May The last settlement of the Etzion block captured by the Arabs Its survivors watched their settlement being plundered

TRANS-JORDAN

River Jordan

Ramallah

Jericho

Bet Haarava

Ataroth

Nebi Samuel

Neve Yaakov

Biddu

Beir Surik

Kolonia

Shuafat

Sheikh Jarrah

Kallia

Neve Ilan

Saris

Motza

Castel

Deir Yassin

Mount Scopus

JERUSALEM

Ein Feshha

Talpioth

Ramat Rachel

Battir

Beit Jalla

Bethlehem

Dead Sea

Revadim

Massuat Yitzhak

Ein Tzurim

Kfar Etzion

0 1 2 3
Miles

Hebron

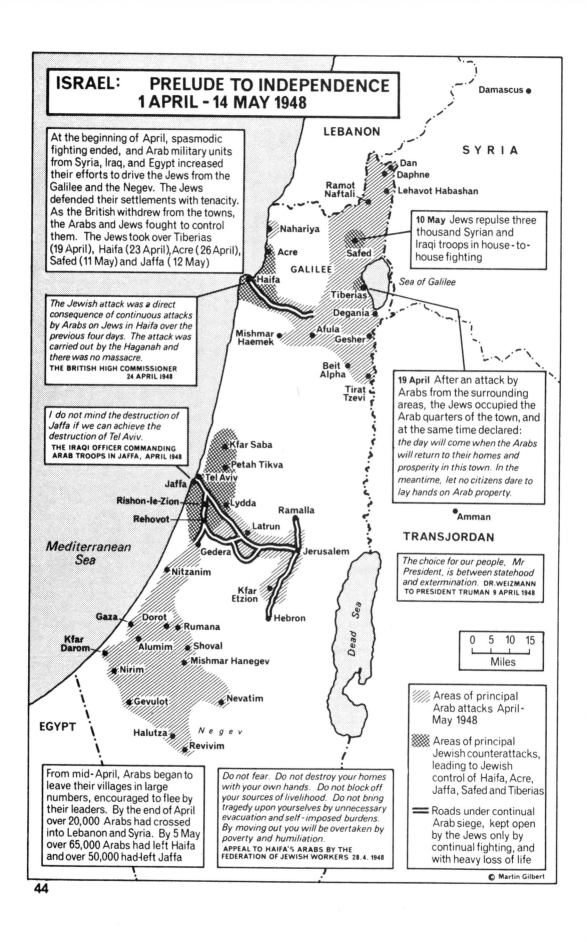

ISRAEL: PRELUDE TO INDEPENDENCE 1 APRIL - 14 MAY 1948

LEBANON

SYRIA

Damascus ●

At the beginning of April, spasmodic fighting ended, and Arab military units from Syria, Iraq, and Egypt increased their efforts to drive the Jews from the Galilee and the Negev. The Jews defended their settlements with tenacity. As the British withdrew from the towns, the Arabs and Jews fought to control them. The Jews took over Tiberias (19 April), Haifa (23 April), Acre (26 April), Safed (11 May) and Jaffa (12 May)

10 May Jews repulse three thousand Syrian and Iraqi troops in house-to-house fighting

Dan
Daphne
Ramot Naftali
Lehavot Habashan
Nahariya
Acre
Safed
GALILEE
Haifa
Sea of Galilee
Tiberias
Degania
Mishmar Haemek
Afula
Gesher
Beit Alpha
Tirat Tzevi

The Jewish attack was a direct consequence of continuous attacks by Arabs on Jews in Haifa over the previous four days. The attack was carried out by the Haganah and there was no massacre.
THE BRITISH HIGH COMMISSIONER 24 APRIL 1948

I do not mind the destruction of Jaffa if we can achieve the destruction of Tel Aviv.
THE IRAQI OFFICER COMMANDING ARAB TROOPS IN JAFFA, APRIL 1948

19 April After an attack by Arabs from the surrounding areas, the Jews occupied the Arab quarters of the town, and at the same time declared: *the day will come when the Arabs will return to their homes and prosperity in this town. In the meantime, let no citizens dare to lay hands on Arab property.*

Kfar Saba
Petah Tikva
Jaffa
Tel Aviv
Rishon-le-Zion
Lydda
Rehovot
Ramalla
Latrun
Gedera
Jerusalem
Nitzanim
Kfar Etzion
Hebron

●Amman

TRANSJORDAN

The choice for our people, Mr President, is between statehood and extermination. **DR.WEIZMANN TO PRESIDENT TRUMAN 9 APRIL 1948**

Mediterranean Sea

Gaza
Dorot
Kfar Darom
Rumana
Alumim
Shoval
Mishmar Hanegev
Nirim
Gevulot
Nevatim
Negev
Halutza
Revivim

Dead Sea

```
0   5   10   15
|__|__|__|__|
     Miles
```

EGYPT

From mid-April, Arabs began to leave their villages in large numbers, encouraged to flee by their leaders. By the end of April over 20,000 Arabs had crossed into Lebanon and Syria. By 5 May over 65,000 Arabs had left Haifa and over 50,000 had left Jaffa

Do not fear. Do not destroy your homes with your own hands. Do not block off your sources of livelihood. Do not bring tragedy upon yourselves by unnecessary evacuation and self-imposed burdens. By moving out you will be overtaken by poverty and humiliation.
APPEAL TO HAIFA'S ARABS BY THE FEDERATION OF JEWISH WORKERS 28.4.1948

///// Areas of principal Arab attacks April-May 1948

▨▨ Areas of principal Jewish counterattacks, leading to Jewish control of Haifa, Acre, Jaffa, Safed and Tiberias

▬▬ Roads under continual Arab siege, kept open by the Jews only by continual fighting, and with heavy loss of life

© Martin Gilbert

44

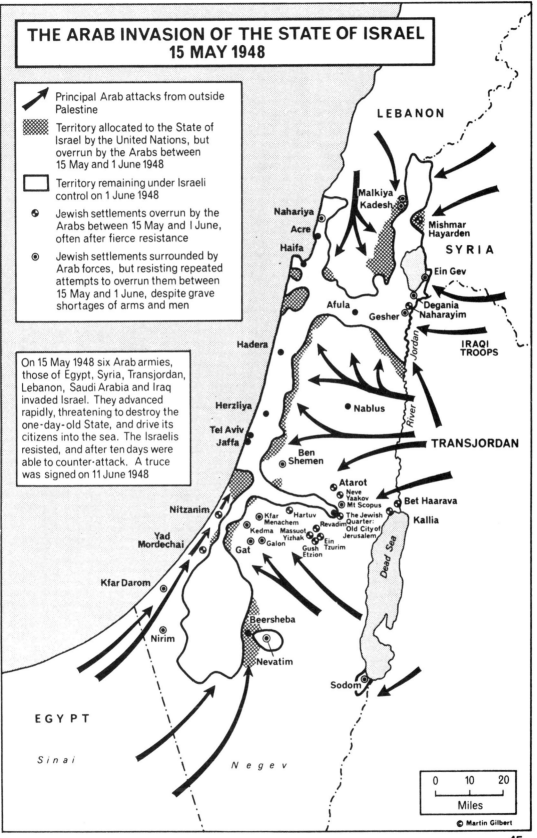

THE ARAB INVASION OF THE STATE OF ISRAEL 15 MAY 1948

↗ Principal Arab attacks from outside Palestine

▨ Territory allocated to the State of Israel by the United Nations, but overrun by the Arabs between 15 May and 1 June 1948

☐ Territory remaining under Israeli control on 1 June 1948

⊙ Jewish settlements overrun by the Arabs between 15 May and I June, often after fierce resistance

◉ Jewish settlements surrounded by Arab forces, but resisting repeated attempts to overrun them between 15 May and 1 June, despite grave shortages of arms and men

On 15 May 1948 six Arab armies, those of Egypt, Syria, Transjordan, Lebanon, Saudi Arabia and Iraq invaded Israel. They advanced rapidly, threatening to destroy the one-day-old State, and drive its citizens into the sea. The Israelis resisted, and after ten days were able to counter-attack. A truce was signed on 11 June 1948

LEBANON

Malkiya
Kadesh

Nahariya
Acre
Haifa

Mishmar Hayarden

SYRIA

Ein Gev

Afula
Gesher

Degania
Naharayim

IRAQI TROOPS

Hadera

Herzliya

Nablus

River Jordan

Tel Aviv
Jaffa

Ben Shemen

TRANSJORDAN

Atarot
Neve Yaakov
Mt Scopus

Nitzanim

Kfar Menachem
Kedma
Massuot
Yizhak
Galon

Hartuv
Revadim
The Jewish Quarter: Old City of Jerusalem
Ein Tzurim
Gush Etzion

Bet Haarava
Kallia

Yad Mordechai

Gat

Dead Sea

Kfar Darom

Nirim

Beersheba

Nevatim

Sodom

EGYPT

Sinai

Negev

0   10   20
Miles

© Martin Gilbert

45

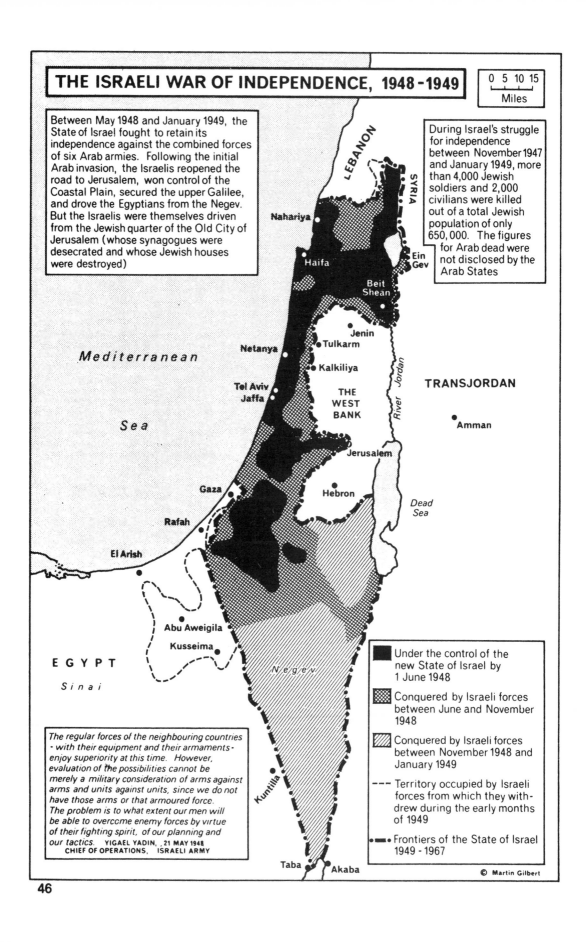

THE ISRAELI WAR OF INDEPENDENCE, 1948-1949

0 5 10 15
Miles

Between May 1948 and January 1949, the State of Israel fought to retain its independence against the combined forces of six Arab armies. Following the initial Arab invasion, the Israelis reopened the road to Jerusalem, won control of the Coastal Plain, secured the upper Galilee, and drove the Egyptians from the Negev. But the Israelis were themselves driven from the Jewish quarter of the Old City of Jerusalem (whose synagogues were desecrated and whose Jewish houses were destroyed)

During Israel's struggle for independence between November 1947 and January 1949, more than 4,000 Jewish soldiers and 2,000 civilians were killed out of a total Jewish population of only 650,000. The figures for Arab dead were not disclosed by the Arab States

LEBANON

SYRIA

Nahariya

Haifa

Ein Gev

Beit Shean

Jenin

Mediterranean

Netanya

Tulkarm

Kalkiliya

Tel Aviv
Jaffa

THE WEST BANK

River Jordan

TRANSJORDAN

Sea

Amman

Jerusalem

Gaza

Hebron

Dead Sea

Rafah

El Arish

Abu Aweigila

Kusseima

E G Y P T

Negev

Sinai

The regular forces of the neighbouring countries - with their equipment and their armaments - enjoy superiority at this time. However, evaluation of the possibilities cannot be merely a military consideration of arms against arms and units against units, since we do not have those arms or that armoured force. The problem is to what extent our men will be able to overcome enemy forces by virtue of their fighting spirit, of our planning and our tactics. YIGAEL YADIN, 21 MAY 1948
CHIEF OF OPERATIONS, ISRAELI ARMY

Kuntilla

Under the control of the new State of Israel by 1 June 1948

Conquered by Israeli forces between June and November 1948

Conquered by Israeli forces between November 1948 and January 1949

Territory occupied by Israeli forces from which they withdrew during the early months of 1949

Frontiers of the State of Israel 1949 - 1967

Taba

Akaba

© Martin Gilbert

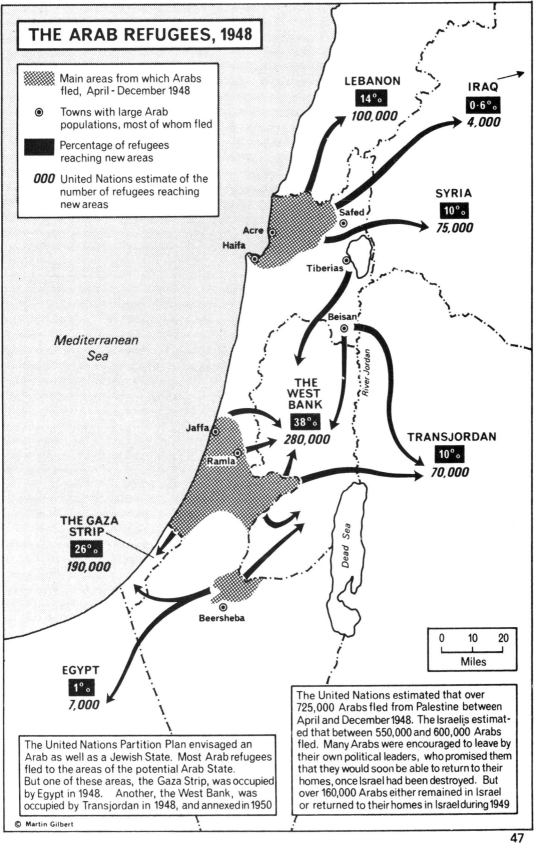

THE ARAB REFUGEES, 1948

Main areas from which Arabs fled, April - December 1948

⊙ Towns with large Arab populations, most of whom fled

■ Percentage of refugees reaching new areas

000 United Nations estimate of the number of refugees reaching new areas

LEBANON 14% *100,000*

IRAQ 0·6% *4,000*

SYRIA 10% *75,000*

Safed ⊙

Acre ⊙
Haifa ⊙

Tiberias ⊙

Beisan ⊙

Mediterranean Sea

THE WEST BANK 38% *280,000*

River Jordan

TRANSJORDAN 10% *70,000*

Jaffa ⊙

Ramla ⊙

THE GAZA STRIP 26% *190,000*

Dead Sea

Beersheba ⊙

0   10   20
Miles

EGYPT 1% *7,000*

The United Nations estimated that over 725,000 Arabs fled from Palestine between April and December 1948. The Israelis estimated that between 550,000 and 600,000 Arabs fled. Many Arabs were encouraged to leave by their own political leaders, who promised them that they would soon be able to return to their homes, once Israel had been destroyed. But over 160,000 Arabs either remained in Israel or returned to their homes in Israel during 1949

The United Nations Partition Plan envisaged an Arab as well as a Jewish State. Most Arab refugees fled to the areas of the potential Arab State. But one of these areas, the Gaza Strip, was occupied by Egypt in 1948. Another, the West Bank, was occupied by Transjordan in 1948, and annexed in 1950

© Martin Gilbert

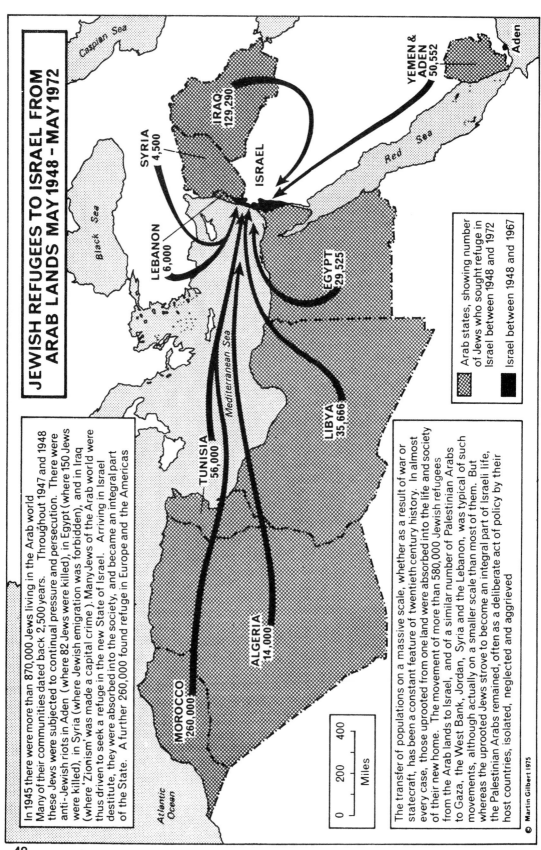

JEWISH REFUGEES TO ISRAEL FROM ARAB LANDS MAY 1948 – MAY 1972

In 1945 there were more than 870,000 Jews living in the Arab world. Many of their communities dated back 2,500 years. Throughout 1947 and 1948 these Jews were subjected to continual pressure and persecution. There were anti-Jewish riots in Aden (where 82 Jews were killed), in Egypt (where 150 Jews were killed), in Syria (where Jewish emigration was forbidden), and in Iraq (where 'Zionism' was made a capital crime). Many Jews of the Arab world were thus driven to seek a refuge in the new State of Israel. Arriving in Israel destitute, they were absorbed into the society, and became an integral part of the State. A further 260,000 found refuge in Europe and the Americas

Caspian Sea

Black Sea

IRAQ
129,290

SYRIA
4,500

ISRAEL

LEBANON
6,000

YEMEN & ADEN
50,552

Red Sea

Aden

Atlantic Ocean

EGYPT
29,525

Mediterranean Sea

TUNISIA
56,000

LIBYA
35,666

MOROCCO
260,000

ALGERIA
14,000

Arab states, showing number of Jews who sought refuge in Israel between 1948 and 1972

Israel between 1948 and 1967

0   200   400
Miles

The transfer of populations on a massive scale, whether as a result of war or statecraft, has been a constant feature of twentieth century history. In almost every case, those uprooted from one land were absorbed into the life and society of their new home. The movement of more than 580,000 Jewish refugees from the Arab lands to Israel, and of a similar number of Palestinian Arabs to Gaza, the West Bank, Jordan, Syria and the Lebanon, was typical of such movements, although actually on a smaller scale than most of them. But whereas the uprooted Jews strove to become an integral part of Israeli life, the Palestinian Arabs remained, often as a deliberate act of policy by their host countries, isolated, neglected and aggrieved

© Martin Gilbert 1975

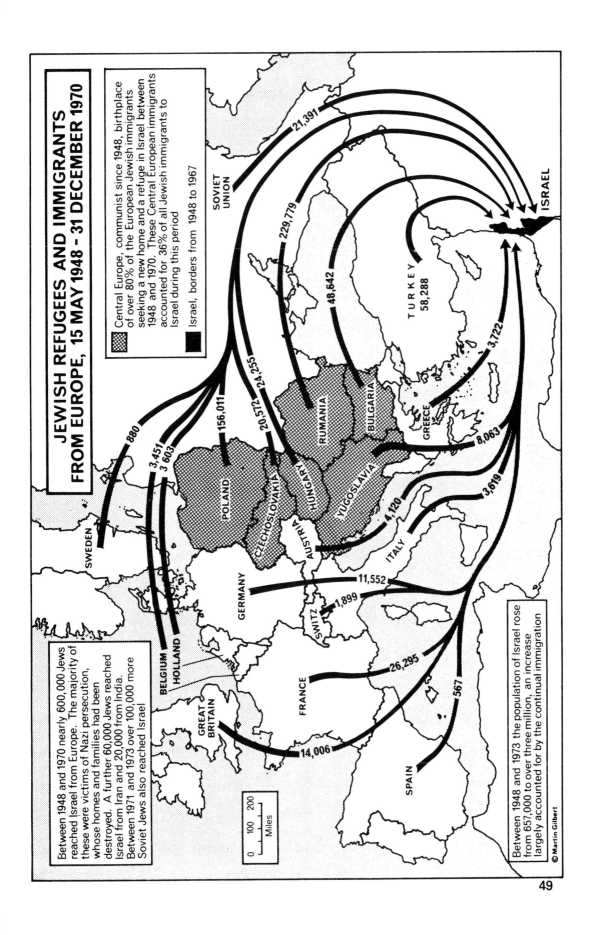

JEWISH REFUGEES AND IMMIGRANTS FROM EUROPE, 15 MAY 1948 – 31 DECEMBER 1970

Between 1948 and 1970 nearly 600,000 Jews reached Israel from Europe. The majority of these were victims of Nazi persecution, whose homes and families had been destroyed. A further 60,000 Jews reached Israel from Iran and 20,000 from India. Between 1971 and 1973 over 100,000 more Soviet Jews also reached Israel

Central Europe, communist since 1948, birthplace of over 80% of the European Jewish immigrants seeking a new home and a refuge in Israel between 1948 and 1970. These Central European immigrants accounted for 36% of all Jewish immigrants to Israel during this period

Israel, borders from 1948 to 1967

0   100   200
Miles

Between 1948 and 1973 the population of Israel rose from 657,000 to over three million, an increase largely accounted for by the continual immigration

SWEDEN 880

SOVIET UNION 21,391

POLAND 156,011

229,779

20,572

24,255

RUMANIA

48,642

BULGARIA

TURKEY 58,288

3,451
3,603
BELGIUM
HOLLAND

CZECHOSLOVAKIA

HUNGARY

AUSTRIA

YUGOSLAVIA

GREECE

8,063

3,722

3,619

4,120

GERMANY

SWITZ 1,899

11,552

ITALY

FRANCE

26,295

SPAIN 567

GREAT BRITAIN

14,006

ISRAEL

© Martin Gilbert

49

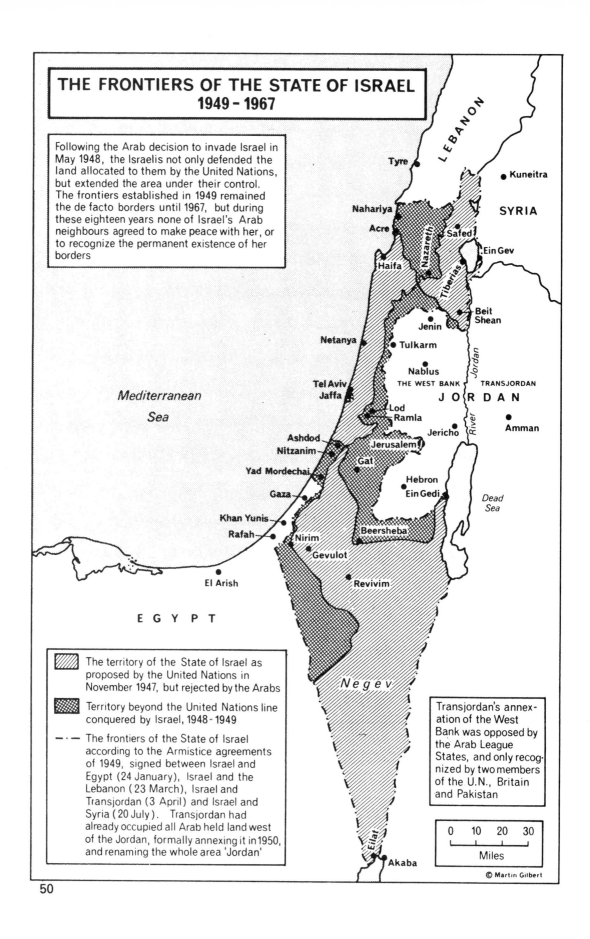

THE FRONTIERS OF THE STATE OF ISRAEL
1949 - 1967

Following the Arab decision to invade Israel in May 1948, the Israelis not only defended the land allocated to them by the United Nations, but extended the area under their control. The frontiers established in 1949 remained the de facto borders until 1967, but during these eighteen years none of Israel's Arab neighbours agreed to make peace with her, or to recognize the permanent existence of her borders

LEBANON

Tyre

Nahariya

Acre

Haifa

● Kuneitra

SYRIA

Nazareth

● Safed

● Ein Gev

Tiberias

Beit Shean

Jenin

Netanya

● Tulkarm

Nablus

Tel Aviv
Jaffa

THE WEST BANK

TRANSJORDAN

J O R D A N

Lod
Ramla

Jerusalem

Jericho

River Jordan

● Amman

Mediterranean
Sea

Ashdod
Nitzanim

Yad Mordechai

Gat

Gaza

Hebron
Ein Gedi

Dead
Sea

Khan Yunis

Rafah

Nirim

Beersheba

Gevulot

El Arish

Revivim

E G Y P T

N e g e v

The territory of the State of Israel as proposed by the United Nations in November 1947, but rejected by the Arabs

Territory beyond the United Nations line conquered by Israel, 1948-1949

–·– The frontiers of the State of Israel according to the Armistice agreements of 1949, signed between Israel and Egypt (24 January), Israel and the Lebanon (23 March), Israel and Transjordan (3 April) and Israel and Syria (20 July). Transjordan had already occupied all Arab held land west of the Jordan, formally annexing it in 1950, and renaming the whole area 'Jordan'

Transjordan's annexation of the West Bank was opposed by the Arab League States, and only recognized by two members of the U.N., Britain and Pakistan

| 0 | 10 | 20 | 30 |

Miles

Eilat

● Akaba

© Martin Gilbert

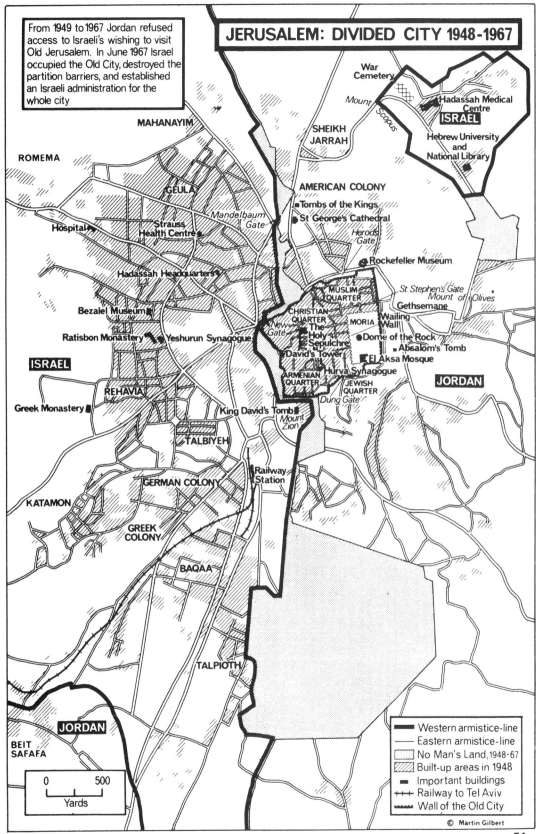

JERUSALEM: DIVIDED CITY 1948-1967

From 1949 to 1967 Jordan refused access to Israeli's wishing to visit Old Jerusalem. In June 1967 Israel occupied the Old City, destroyed the partition barriers, and established an Israeli administration for the whole city

War Cemetery

Mount Scopus

ISRAEL

Hadassah Medical Centre

Hebrew University and National Library

MAHANAYIM

SHEIKH JARRAH

ROMEMA

GEULA

AMERICAN COLONY

Tombs of the Kings

St George's Cathedral

Mandelbaum Gate

Hospital

Strauss Health Centre

Herods Gate

Rockefeller Museum

Hadassah Headquarters

MUSLIM QUARTER

St Stephen's Gate

Mount of Olives

Bezalel Museum

CHRISTIAN QUARTER

Gethsemane

MORIA

Wailing Wall

Ratisbon Monastery

Yeshurun Synagogue

New Gate

The Holy Sepulchre

Dome of the Rock

Absalom's Tomb

ISRAEL

David's Tower

Hurva Synagogue

El Aksa Mosque

JORDAN

REHAVIA

ARMENIAN QUARTER

JEWISH QUARTER

Greek Monastery

Dung Gate

King David's Tomb

Mount Zion

TALBIYEH

Railway Station

GERMAN COLONY

KATAMON

GREEK COLONY

BAQAA

TALPIOTH

JORDAN

BEIT SAFAFA

0   500

Yards

	Western armistice-line
	Eastern armistice-line
	No Man's Land,1948-67
	Built-up areas in 1948
	Important buildings
	Railway to Tel Aviv
	Wall of the Old City

© Martin Gilbert

51

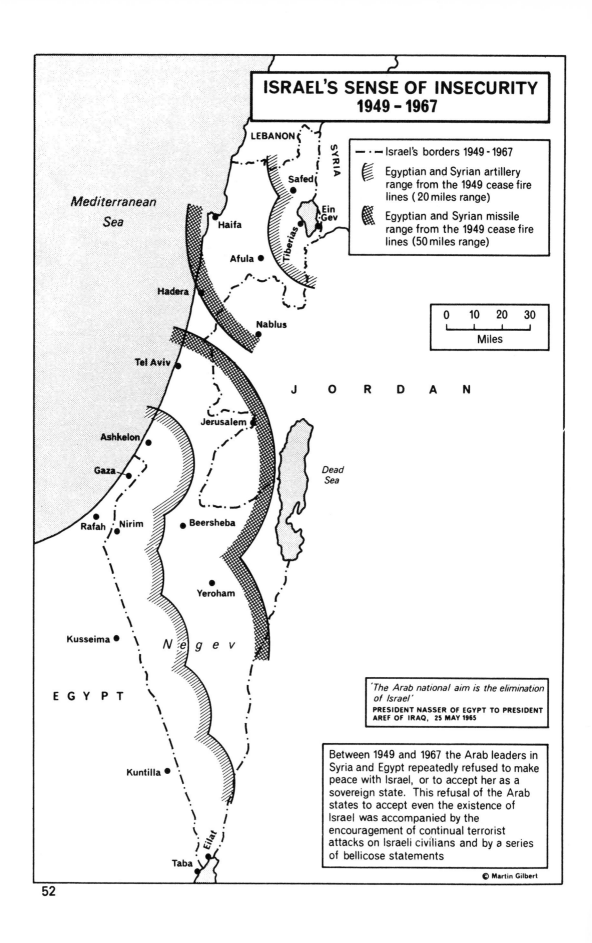

ISRAEL'S SENSE OF INSECURITY 1949 - 1967

Mediterranean Sea

LEBANON

SYRIA

Safed

Ein Gev

Haifa

Tiberias

Afula

Hadera

Nablus

Tel Aviv

J O R D A N

Jerusalem

Ashkelon

Dead Sea

Gaza

Rafah • Nirim

Beersheba

Kusseima

Yeroham

N e g e v

E G Y P T

Kuntilla

Eilat

Taba

Legend:

— · — Israel's borders 1949 - 1967

Egyptian and Syrian artillery range from the 1949 cease fire lines (20 miles range)

Egyptian and Syrian missile range from the 1949 cease fire lines (50 miles range)

0 10 20 30
Miles

'The Arab national aim is the elimination of Israel'
PRESIDENT NASSER OF EGYPT TO PRESIDENT AREF OF IRAQ, 25 MAY 1965

Between 1949 and 1967 the Arab leaders in Syria and Egypt repeatedly refused to make peace with Israel, or to accept her as a sovereign state. This refusal of the Arab states to accept even the existence of Israel was accompanied by the encouragement of continual terrorist attacks on Israeli civilians and by a series of bellicose statements

© Martin Gilbert

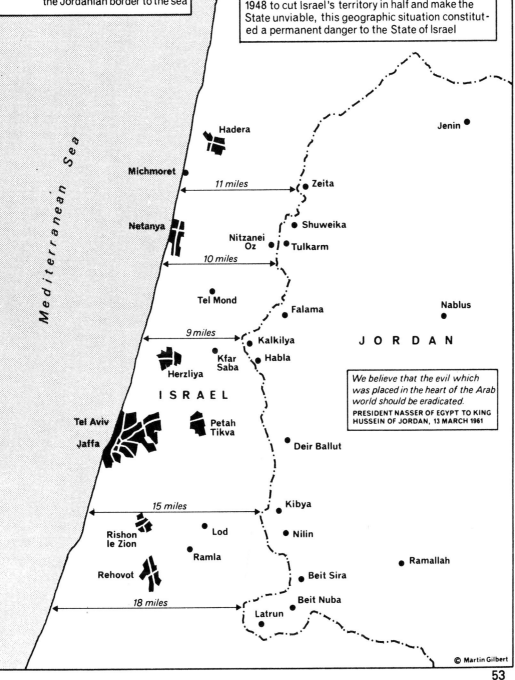

CENTRAL ISRAEL AND THE JORDAN BORDER 1949–1967

0 1 2 3 4 5
Miles

–·–·– The cease-fire line between Israel and Jordan, 1949-1967

Principal built-up areas

Distances across Israel from the Jordanian border to the sea

Between 1949 and 1967 the whole of central Israel, from Hadera to Rehovot, lay in a narrow belt of land sandwiched between Jordan and the sea. At its most narrow, Israel was only nine miles wide, and all Israeli territory shown on this map was within Jordanian artillery range. In view of Arab threats in 1948 to cut Israel's territory in half and make the State unviable, this geographic situation constituted a permanent danger to the State of Israel

Mediterranean Sea

Jenin

Hadera

Michmoret
11 miles — Zeita

Netanya
Shuweika
Nitzanei Oz — Tulkarm

10 miles

Tel Mond
Falama
Nablus

9 miles
Kalkilya
Kfar Saba — Habla

J O R D A N

Herzliya

I S R A E L

We believe that the evil which was placed in the heart of the Arab world should be eradicated.
PRESIDENT NASSER OF EGYPT TO KING HUSSEIN OF JORDAN, 13 MARCH 1961

Tel Aviv — Petah Tikva

Jaffa

Deir Ballut

15 miles — Kibya

Rishon le Zion
Lod
Nilin

Ramla
Ramallah

Rehovot
Beit Sira

Beit Nuba
18 miles — Latrun

© Martin Gilbert

53

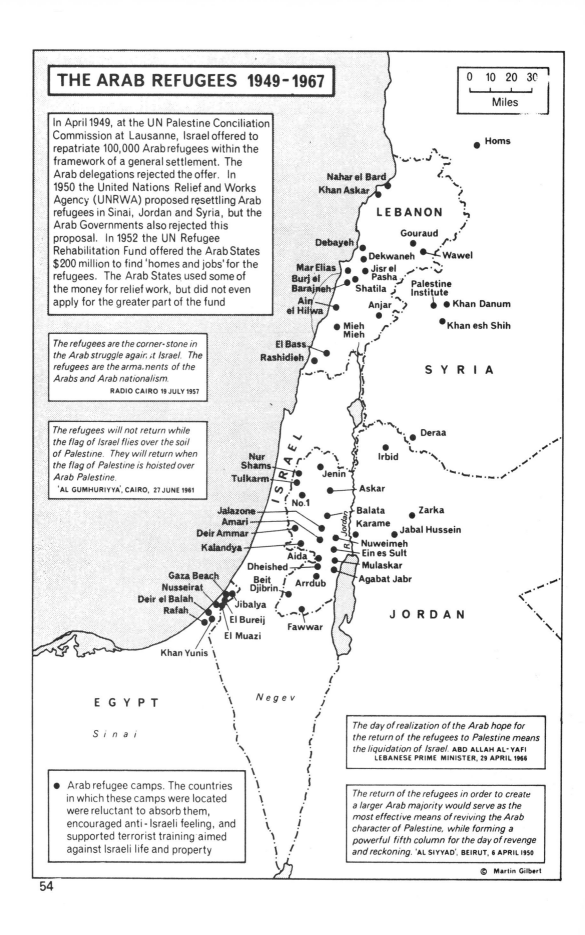

THE ARAB REFUGEES 1949-1967

0 10 20 30
Miles

In April 1949, at the UN Palestine Conciliation Commission at Lausanne, Israel offered to repatriate 100,000 Arab refugees within the framework of a general settlement. The Arab delegations rejected the offer. In 1950 the United Nations Relief and Works Agency (UNRWA) proposed resettling Arab refugees in Sinai, Jordan and Syria, but the Arab Governments also rejected this proposal. In 1952 the UN Refugee Rehabilitation Fund offered the Arab States $200 million to find 'homes and jobs' for the refugees. The Arab States used some of the money for relief work, but did not even apply for the greater part of the fund

The refugees are the corner-stone in the Arab struggle against Israel. The refugees are the armaments of the Arabs and Arab nationalism.
RADIO CAIRO 19 JULY 1957

The refugees will not return while the flag of Israel flies over the soil of Palestine. They will return when the flag of Palestine is hoisted over Arab Palestine.
'AL GUMHURIYYA', CAIRO, 27 JUNE 1961

Homs

Nahar el Bard
Khan Askar

LEBANON

Gouraud

Debayeh

Dekwaneh
Wawel

Mar Elias
Burj el
Barajneh
Jisr el
Pasha
Shatila
Palestine
Institute

Ain
el Hilwa
Anjar
Khan Danum

Mieh
Mieh
Khan esh Shih

El Bass
Rashidieh

S Y R I A

Deraa

Irbid

Nur
Shams
Jenin
Tulkarm

Askar

No.1

Balata
Zarka
Jalazone
Karame
Amari
Jabal Hussein
Deir Ammar
Nuweimeh
Kalandya
Ein es Sult
Aida
Mulaskar
Dheished
Agabat Jabr
Beit
Gaza Beach
Djibrin
Arrdub
Nusseirat
Deir el Balah
Rafah
Jibalya
J O R D A N
El Bureij
Fawwar
El Muazi

Khan Yunis

E G Y P T

Negev

Sinai

The day of realization of the Arab hope for the return of the refugees to Palestine means the liquidation of Israel. **ABD ALLAH AL-YAFI LEBANESE PRIME MINISTER, 29 APRIL 1966**

● Arab refugee camps. The countries in which these camps were located were reluctant to absorb them, encouraged anti-Israeli feeling, and supported terrorist training aimed against Israeli life and property

The return of the refugees in order to create a larger Arab majority would serve as the most effective means of reviving the Arab character of Palestine, while forming a powerful fifth column for the day of revenge and reckoning. 'AL SIYYAD', BEIRUT, 6 APRIL 1950

© Martin Gilbert

54

THE ARABS OF NORTHERN ISRAEL

LEBANON

◉ Arab and Druze villages in Israel

✪ Towns with mixed Arab and Jewish population

▮ Entirely Arab villages with more than 5,000 inhabitants

Jurdeih

Rihaniya

1949 CEASE-FIRE LINE

Jish

Tuba

1949 CEASE-FIRE LINE

SYRIA

Acre

Rama

Maghar

Sakhnin

Arraba

Tamra

Sea of Galilee

Shefar Am

Haifa

Kafr Kanna

Daliyat el Carmel

Nazareth

JORDAN

Fureidis

Sandala

Umm el Fahm

1949 CEASE-FIRE LINE

Baka

River Jordan

Mediterranean Sea

In 1949 some 160,000 Arabs remained in Israel. By 1966 their numbers had increased to 312,000, or 11% of the total population. Less than 6,000 Arabs left Israel during this period. From 1959, Arabs could join the Israeli Trade Union Organization, and on 1 December 1966 Israeli Military Government was abolished in all Arab areas. Since 1949, Arabs voted in all Israeli elections, and sent their own members to the Israeli Parliament. Since May 1948, both Hebrew and Arabic have been the official languages of the State of Israel, and three Arabic-language daily newspapers have been in regular production. But throughout this period, the Arabs maintained separate communities and cultural life, as did the 30,000 Bedouin who lived in the Negev and around Beersheba

al-Taiyiba

Tira

Jaljulya

Jaffa

The majority of the Arabs who remained in Israel after May 1948 lived in the northernmost part of the State. Whereas the Arabs who fled from Israel in 1948 were for the most part confined to refugee camps by their fellow Arab hosts and deliberately cut off from the economic development of the States in which they lived, the Arabs of Israel continued to live unmolested in their original homes, gained materially from Israel's own material successes, and received direct Israeli aid for irrigation, reclamation, mechanization and social welfare (including education, health and housing)

Lod

Karyat Jawarish

Ramla

0 5
Miles

© Martin Gilbert

55

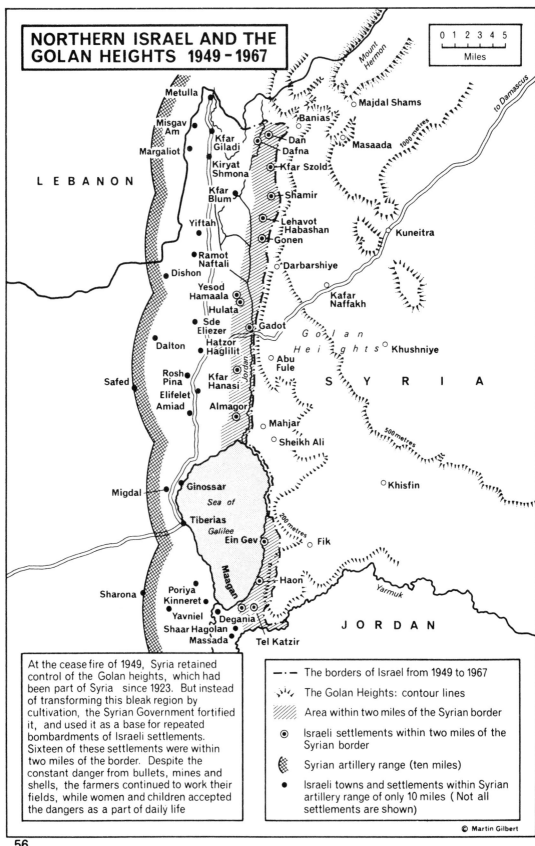

NORTHERN ISRAEL AND THE GOLAN HEIGHTS 1949-1967

0 1 2 3 4 5
Miles

Mount Hermon

to Damascus

Metulla

Misgav Am

Margaliot

Majdal Shams

Banias

LEBANON

Kfar Giladi

Dan
Dafna

Masaada

1000 metres

Kiryat Shmona

Kfar Szold

Kfar Blum

Shamir

Yiftah

Lehavot Habashan
Gonen

Kuneitra

Ramot Naftali

Darbarshiye

Dishon

Yesod Hamaala

Kafar Naffakh

Hulata

Sde Eliezer

Gadot

G o l a n

H e i g h t s

Khushniye

Dalton

Hatzor Haglilit

Abu Fule

S Y R I A

Rosh Pina

Kfar Hanasi

Safed

Elifelet

Amiad

Almagor

500 metres

Mahjar

Sheikh Ali

Migdal

Ginossar

Sea of

Khisfin

Tiberias

Galilee

200 metres

Ein Gev

Fik

Maagan

Haon

Yarmuk

Sharona

Poriya
Kinneret

Yavniel

Degania

JORDAN

Shaar Hagolan

Massada

Tel Katzir

At the cease fire of 1949, Syria retained control of the Golan heights, which had been part of Syria since 1923. But instead of transforming this bleak region by cultivation, the Syrian Government fortified it, and used it as a base for repeated bombardments of Israeli settlements. Sixteen of these settlements were within two miles of the border. Despite the constant danger from bullets, mines and shells, the farmers continued to work their fields, while women and children accepted the dangers as a part of daily life

—·— The borders of Israel from 1949 to 1967

The Golan Heights: contour lines

Area within two miles of the Syrian border

⊙ Israeli settlements within two miles of the Syrian border

Syrian artillery range (ten miles)

● Israeli towns and settlements within Syrian artillery range of only 10 miles (Not all settlements are shown)

© Martin Gilbert

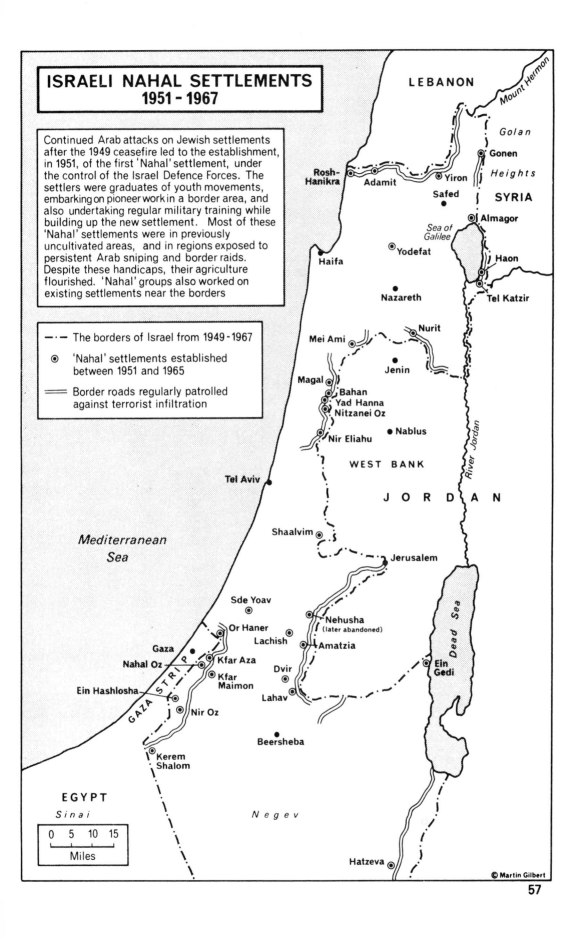

ISRAELI NAHAL SETTLEMENTS
1951 - 1967

Continued Arab attacks on Jewish settlements
after the 1949 ceasefire led to the establishment,
in 1951, of the first 'Nahal' settlement, under
the control of the Israel Defence Forces. The
settlers were graduates of youth movements,
embarking on pioneer work in a border area, and
also undertaking regular military training while
building up the new settlement. Most of these
'Nahal' settlements were in previously
uncultivated areas, and in regions exposed to
persistent Arab sniping and border raids.
Despite these handicaps, their agriculture
flourished. 'Nahal' groups also worked on
existing settlements near the borders

—·— The borders of Israel from 1949 - 1967

⊙ 'Nahal' settlements established
between 1951 and 1965

══ Border roads regularly patrolled
against terrorist infiltration

LEBANON

Mount Hermon

Golan

Gonen

Heights

Rosh-
Hanikra
Adamit
Yiron
Safed
SYRIA

Sea of
Galilee
Almagor

Haifa
Yodefat
Haon

Nazareth
Tel Katzir

Mei Ami
Nurit

Jenin

Magal
Bahan
Yad Hanna
Nitzanei Oz
Nir Eliahu
Nablus

WEST BANK

River Jordan

JORDAN

Tel Aviv

Mediterranean
Sea

Shaalvim

Jerusalem

Sde Yoav

Nehusha
(later abandoned)

Or Haner
Lachish
Amatzia

Gaza
Kfar Aza
Nahal Oz
Dvir
Kfar
Maimon
Ein Hashlosha
Lahav

Nir Oz

Dead Sea

Ein
Gedi

Beersheba

Kerem
Shalom

EGYPT
Sinai
Negev

| 0 | 5 | 10 | 15 |
Miles

Hatzeva

© Martin Gilbert

57

TERRORIST RAIDS INTO ISRAEL 1951 - 1956

Palestinian terrorist groups, or Fedayeen, began systematic raids into Israel from 1950. Towards the end of 1954, the Egyptian Government supervised the formal establishment of Palestinian terrorist groups in the Gaza strip and north-eastern Sinai. Throughout 1955 an increasing number of raids were launched into Israel. From 1951 to 1956, Israeli vehicles were ambushed, farms attacked, fields boobytrapped and roads mined. Fedayeen from Gaza also infiltrated into Jordan, and operated from there. Saudi Arabia, Syria and Lebanon each gave the Fedayeen support and refuge. Local Jordanian-Palestinian Fedayeen were also active operating from the West Bank

ISRAELI DEATHS AS A RESULT OF FEDAYEEN ATTACKS

YEAR	FROM	ISRAELI DEAD
1951	JORDAN	111
	EGYPT	26
1952	JORDAN	114
	EGYPT	48
1953	JORDAN	124
	EGYPT	38
1954	JORDAN	117
	EGYPT	50
1955	JORDAN	37
	EGYPT	241
1951-55	SYRIA	55
	LEBANON	6

◉ Centres of anti-Israel activity

⇗ Moral and material support for Fedayeen attacks

➤ Movement of Fedayeen groups

▨ Areas of Fedayeen activity against Israel. With Egyptian encouragement, the Fedayeen also incited demonstrations inside Jordan against the Jordanian regime

0 10 20 30 40 50
Miles

Between 1951 and 1955 967 Israelis were killed by Arab terrorists operating inside Israel's 1949 borders

© Martin Gilbert

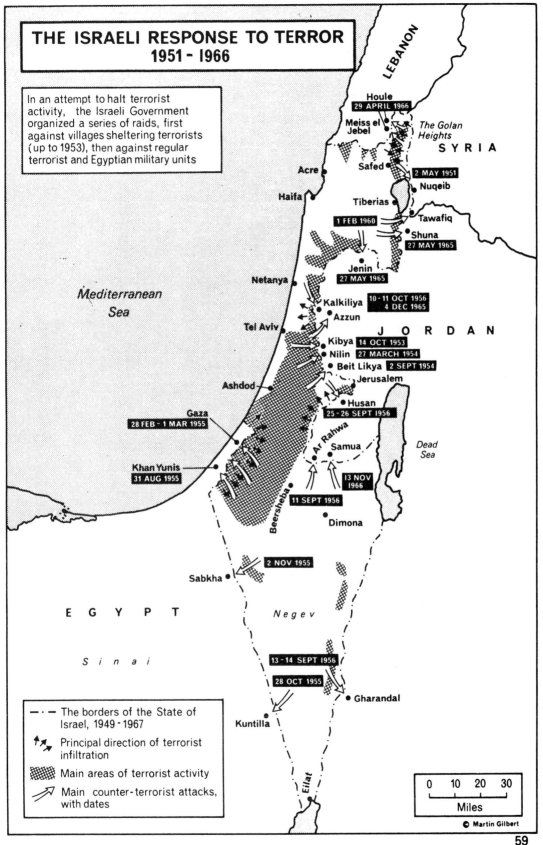

THE ISRAELI RESPONSE TO TERROR 1951 - 1966

In an attempt to halt terrorist activity, the Israeli Government organized a series of raids, first against villages sheltering terrorists (up to 1953), then against regular terrorist and Egyptian military units

LEBANON

Houle
29 APRIL 1966

Meiss el Jebel

The Golan Heights

SYRIA

Acre

Safed

2 MAY 1951

Nuqeib

Haifa

Tiberias

Tawafiq

1 FEB 1960

Shuna
27 MAY 1965

Mediterranean Sea

Netanya

Jenin
27 MAY 1965

Kalkiliya

Azzun

J O R D A N

Tel Aviv

10 - 11 OCT 1956
4 DEC 1965

Kibya   14 OCT 1953

Nilin   27 MARCH 1954

Beit Likya   2 SEPT 1954

Ashdod

Jerusalem

Husan
25 - 26 SEPT 1956

Gaza
28 FEB - 1 MAR 1955

Ar Rahwa

Samua

Dead Sea

Khan Yunis
31 AUG 1955

13 NOV 1966

Beersheba

11 SEPT 1956

Dimona

2 NOV 1955

Sabkha

E G Y P T

Negev

Sinai

13 - 14 SEPT 1956

28 OCT 1955

Gharandal

Kuntilla

Eilat

—·— The borders of the State of Israel, 1949 - 1967

↑↗ Principal direction of terrorist infiltration

▨ Main areas of terrorist activity

↗ Main counter-terrorist attacks, with dates

0   10   20   30
Miles

© Martin Gilbert

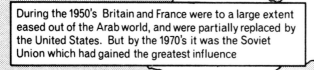

THE CHANGING BALANCE OF POWER IN THE ARAB WORLD
1953 – 1973

During the 1950's Britain and France were to a large extent eased out of the Arab world, and were partially replaced by the United States. But by the 1970's it was the Soviet Union which had gained the greatest influence

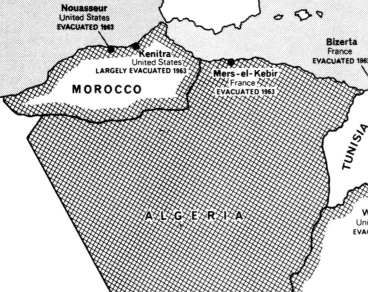

Nouasseur
United States
EVACUATED 1963

Kenitra
United States
LARGELY EVACUATED 1963

MOROCCO

Mers-el-Kebir
France
EVACUATED 1963

Bizerta
France
EVACUATED 1963

TUNISIA

Mediterranean Sea

El-Adhem
British
EVACUATED 1970

Wheelus
United States
EVACUATED 1970

A L G E R I A

Tobruk
British
EVACUATED 1970

L I B Y A

British and French influence in the Arab world was greatly weakened in 1956, after British troops landed at Port Said, in an unsuccessful attempt to reverse, by force, Egypt's nationalisation of the Suez Canal

```
0   100  200  300  400
Miles
```

- ● Western naval and air bases (French, British and United States) in existence in 1953, but abandoned by 1973

- ◉ Soviet naval bases and facilities established between 1963 and 1973

- ☀ British air base (in Cyprus) whose use was denied to the United States by Britain during the Middle East war of October 1973

- ▨ Arab countries which received the majority of their arms from the Soviet Union, 1971-1973

- ▨ Other Arab countries receiving Soviet military aid since 1967

© Martin Gilbert

VALUE OF SOVIET ARMS SUPPLIES 1965-1970 (in dollars)	
Egypt	$4,500 million
Iraq	$500 million
Syria	$450 million
Algeria	$250 million
Others (Sudan, South Yemen, Yemen, Libya)	$1,000 million
Total	$6,700 million

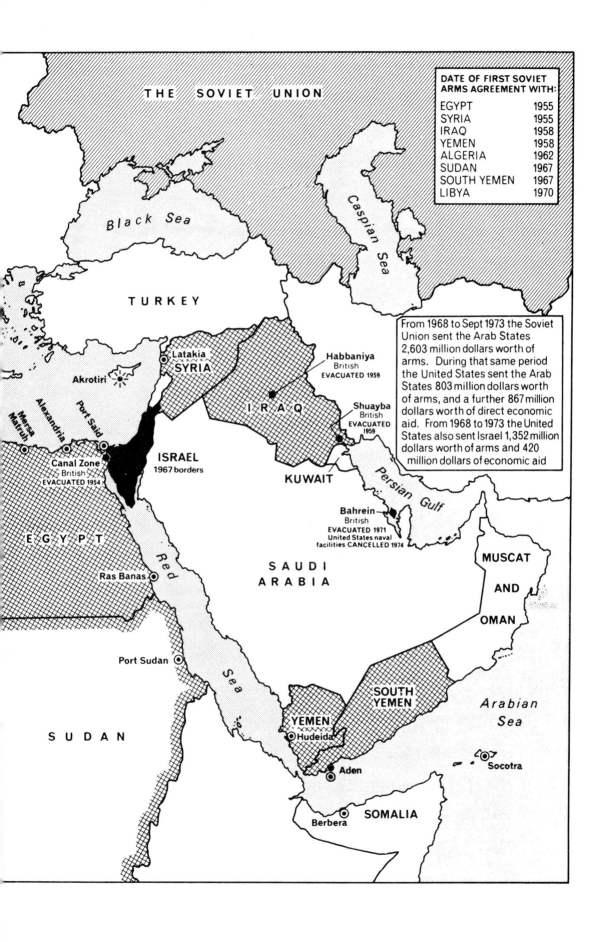

THE SOVIET UNION

DATE OF FIRST SOVIET
ARMS AGREEMENT WITH:

EGYPT   1955
SYRIA   1955
IRAQ   1958
YEMEN   1958
ALGERIA   1962
SUDAN   1967
SOUTH YEMEN   1967
LIBYA   1970

Black Sea

Caspian Sea

TURKEY

Latakia
SYRIA
Akrotiri

Habbaniya
British
EVACUATED 1959

I R A Q

Shuayba
British
EVACUATED
1959

Alexandria
Mersa
Matruh
Port Said
Canal Zone
British
EVACUATED 1954

ISRAEL
1967 borders

KUWAIT

Persian Gulf

From 1968 to Sept 1973 the Soviet
Union sent the Arab States
2,603 million dollars worth of
arms. During that same period
the United States sent the Arab
States 803 million dollars worth
of arms, and a further 867 million
dollars worth of direct economic
aid. From 1968 to 1973 the United
States also sent Israel 1,352 million
dollars worth of arms and 420
million dollars of economic aid

E G Y P T

Bahrein
British
EVACUATED 1971
United States naval
facilities CANCELLED 1974

S A U D I
A R A B I A

MUSCAT

AND

OMAN

Ras Banas

Red

Port Sudan

Sea

SOUTH
YEMEN

Arabian
Sea

SUDAN

YEMEN
Hudeida

Socotra

Aden

SOMALIA

Berbera

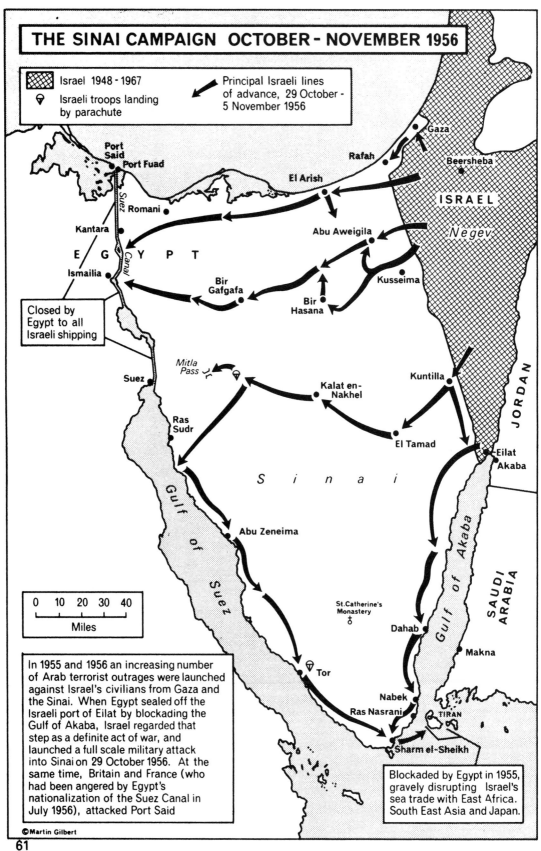

THE SINAI CAMPAIGN OCTOBER – NOVEMBER 1956

Israel 1948 - 1967

Israeli troops landing by parachute

Principal Israeli lines of advance, 29 October - 5 November 1956

Gaza

Port Said
Port Fuad

Rafah

Beersheba

El Arish

ISRAEL

Romani

Kantara

Abu Aweigila

E G Y P T

Negev

Ismailia

Bir Gafgafa

Kusseima

Closed by Egypt to all Israeli shipping

Bir Hasana

Mitla Pass

Suez

Kalat en-Nakhel

Kuntilla

JORDAN

Ras Sudr

El Tamad

Eilat
Akaba

S i n a i

0 10 20 30 40

Miles

Abu Zeneima

St.Catherine's Monastery

Dahab

Makna

SAUDI ARABIA

Gulf of Suez

Gulf of Akaba

Tor

Nabek

Ras Nasrani

TIRAN

In 1955 and 1956 an increasing number of Arab terrorist outrages were launched against Israel's civilians from Gaza and the Sinai. When Egypt sealed off the Israeli port of Eilat by blockading the Gulf of Akaba, Israel regarded that step as a definite act of war, and launched a full scale military attack into Sinai on 29 October 1956. At the same time, Britain and France (who had been angered by Egypt's nationalization of the Suez Canal in July 1956), attacked Port Said

Sharm el-Sheikh

Blockaded by Egypt in 1955, gravely disrupting Israel's sea trade with East Africa. South East Asia and Japan.

©Martin Gilbert

61

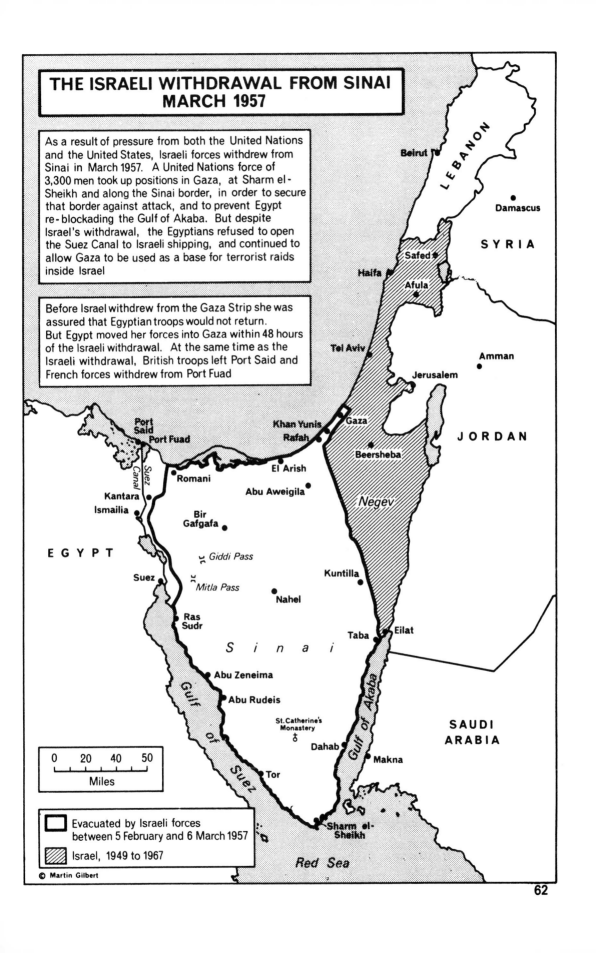

THE ISRAELI WITHDRAWAL FROM SINAI MARCH 1957

As a result of pressure from both the United Nations and the United States, Israeli forces withdrew from Sinai in March 1957. A United Nations force of 3,300 men took up positions in Gaza, at Sharm el-Sheikh and along the Sinai border, in order to secure that border against attack, and to prevent Egypt re-blockading the Gulf of Akaba. But despite Israel's withdrawal, the Egyptians refused to open the Suez Canal to Israeli shipping, and continued to allow Gaza to be used as a base for terrorist raids inside Israel

Before Israel withdrew from the Gaza Strip she was assured that Egyptian troops would not return. But Egypt moved her forces into Gaza within 48 hours of the Israeli withdrawal. At the same time as the Israeli withdrawal, British troops left Port Said and French forces withdrew from Port Fuad

LEBANON

Beirut

Damascus

SYRIA

Safed

Haifa

Afula

Tel Aviv

Amman

Jerusalem

JORDAN

Khan Yunis
Rafah

Gaza

Beersheba

Port
Said
Port Fuad

El Arish

Romani

Abu Aweigila

Negev

Kantara

Ismailia

Bir
Gafgafa

EGYPT

⊃⊂ Giddi Pass

Kuntilla

⊃⊂ Mitla Pass

Suez

Nahel

S i n a i

Ras
Sudr

Taba

Eilat

Abu Zeneima

Abu Rudeis

St. Catherine's
Monastery
⚓

SAUDI
ARABIA

Dahab

Makna

Gulf of Akaba

Gulf of Suez

Tor

Sharm el-
Sheikh

Red Sea

| 0 | 20 | 40 | 50 |

Miles

☐ Evacuated by Israeli forces
between 5 February and 6 March 1957

▨ Israel, 1949 to 1967

© Martin Gilbert

SYRIAN ACTIVITY AGAINST ISRAELI SETTLEMENTS FEBRUARY - OCTOBER 1966

0 1 2 3 4 5
Miles

S Y R I A

L E B A N O N

Shear Yashuv

6 September 7 land reclamation officers wounded by a mine

30 April 4 Israeli workers wounded by Syrian machine gun fire
5 June Syrians shell workers in the fields
6 June Syrian shells set fire to fields

Ashmura
Hulata
Gadot

13 February Syrians shoot at Israeli tractors with mortars. Israeli forces succeed in putting two Syrian tanks out of action

Mahanayim

12 July A tractor driver seriously wounded by a Syrian mine

The border between Syria and Israel was the scene of repeated Syrian bombardment, sniping and minelaying between 1948 and 1967. From their fortifications on the Golan heights, the Syrians tried to disrupt the daily life of the Israeli farmers and fishermen.

Almagor

12 July An afforestation officer killed by a Syrian mine

26 September Syrians fire on a fishing boat

In 1963, at the instigation of the Arab League, Syria tried to divert the headwaters of the Jordan, so that Israel would lose her main source of freshwater, and be unable to complete her plans to harness the Jordan waters for irrigation works throughout Israel. In 1964 Israeli artillery destroyed the Syrian earth-digging equipment which was about to begin the diversion, and Syrian plans were abandoned

Sea
of
Galilee

River Jordan

21 February Syrians fire mortar shells at a patrol escorting fisherman on lake
15 August Syrians open fire on patrol boat. Israelis retaliate. 5 Israeli soldiers wounded. Two Syrian planes shot down

We shall never call for nor accept peace. We shall only accept war. We have resolved to drench this land with your blood, to oust you aggressors, to throw you into the sea   HAFIZ ASSAD, THEN **SYRIAN DEFENCE MINISTER, 24 MAY 1966**

Haon

I S R A E L

Shaar Hagolan

22 February Syrians fire on a tractor
29 March Tractor driver wounded by Syrian machine gun fire
30 March A second tractor driver wounded by Syrian artillery fire
22 October Tractor driver fired at

Jordan
River

J O R D A N

Our Army will be satisfied with nothing less than the disappearance of Israel   SALAH JADID, SYRIAN CHIEF OF STAFF 30 OCTOBER 1964

9 October 4 border policemen killed by a Syrian mine

© Martin Gilbert

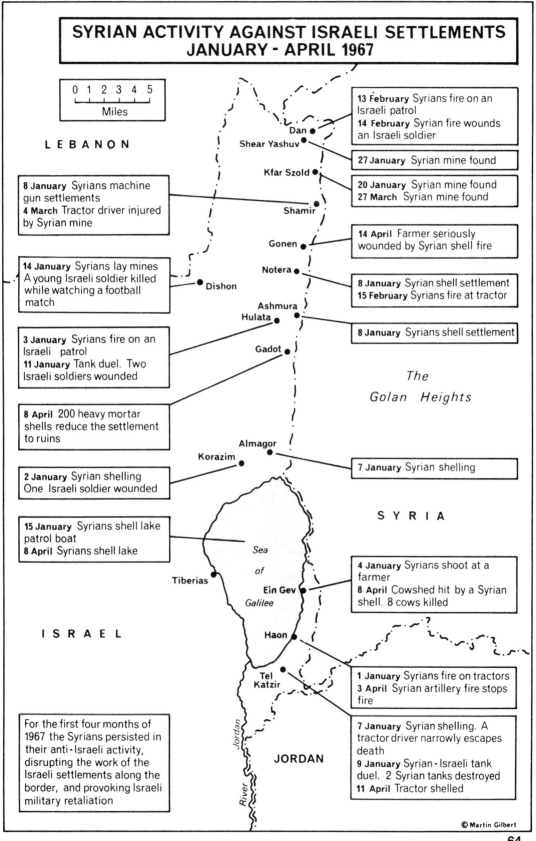

SYRIAN ACTIVITY AGAINST ISRAELI SETTLEMENTS
JANUARY - APRIL 1967

0 1 2 3 4 5
Miles

LEBANON

13 February Syrians fire on an Israeli patrol
14 February Syrian fire wounds an Israeli soldier

27 January Syrian mine found

8 January Syrians machine gun settlements
4 March Tractor driver injured by Syrian mine

20 January Syrian mine found
27 March Syrian mine found

14 April Farmer seriously wounded by Syrian shell fire

14 January Syrians lay mines A young Israeli soldier killed while watching a football match

8 January Syrian shell settlement
15 February Syrians fire at tractor

3 January Syrians fire on an Israeli patrol
11 January Tank duel. Two Israeli soldiers wounded

8 January Syrians shell settlement

8 April 200 heavy mortar shells reduce the settlement to ruins

The Golan Heights

2 January Syrian shelling One Israeli soldier wounded

7 January Syrian shelling

15 January Syrians shell lake patrol boat
8 April Syrians shell lake

S Y R I A

4 January Syrians shoot at a farmer
8 April Cowshed hit by a Syrian shell. 8 cows killed

Sea of Galilee

Dan
Shear Yashuv
Kfar Szold
Shamir
Gonen
Notera
Dishon
Ashmura
Hulata
Gadot
Almagor
Korazim
Tiberias
Ein Gev
Haon
Tel Katzir

I S R A E L

Jordan River

JORDAN

1 January Syrians fire on tractors
3 April Syrian artillery fire stops fire

For the first four months of 1967 the Syrians persisted in their anti-Israeli activity, disrupting the work of the Israeli settlements along the border, and provoking Israeli military retaliation

7 January Syrian shelling. A tractor driver narrowly escapes death
9 January Syrian-Israeli tank duel. 2 Syrian tanks destroyed
11 April Tractor shelled

© Martin Gilbert

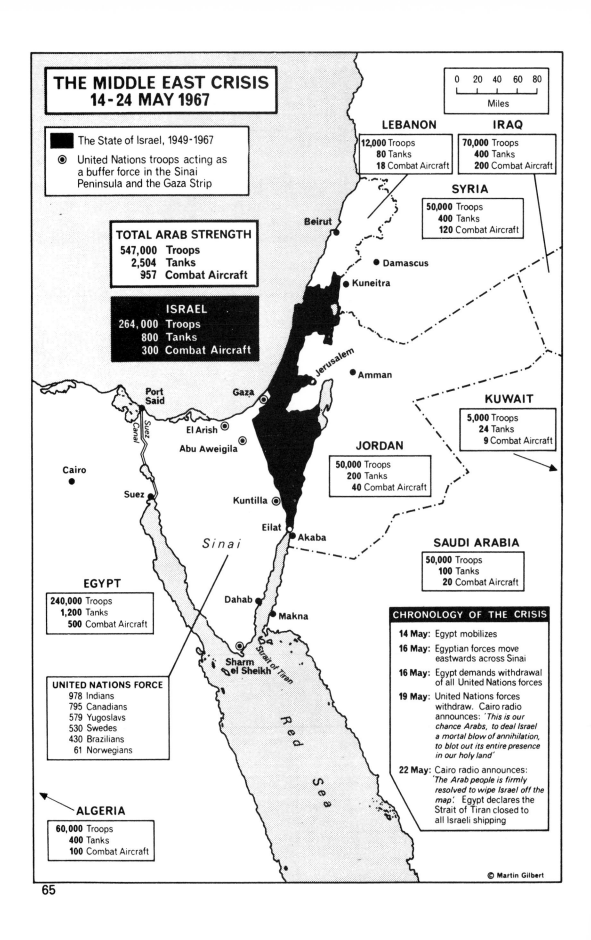

THE MIDDLE EAST CRISIS
14-24 MAY 1967

■ The State of Israel, 1949-1967

◉ United Nations troops acting as a buffer force in the Sinai Peninsula and the Gaza Strip

0 20 40 60 80
Miles

TOTAL ARAB STRENGTH
547,000 Troops
2,504 Tanks
957 Combat Aircraft

ISRAEL
264,000 Troops
800 Tanks
300 Combat Aircraft

LEBANON
12,000 Troops
80 Tanks
18 Combat Aircraft

IRAQ
70,000 Troops
400 Tanks
200 Combat Aircraft

SYRIA
50,000 Troops
400 Tanks
120 Combat Aircraft

Beirut
Damascus
Kuneitra
Jerusalem
Amman

KUWAIT
5,000 Troops
24 Tanks
9 Combat Aircraft

Port Said
Gaza
El Arish
Abu Aweigila
Cairo
Suez

JORDAN
50,000 Troops
200 Tanks
40 Combat Aircraft

Kuntilla
Eilat
Akaba

SAUDI ARABIA
50,000 Troops
100 Tanks
20 Combat Aircraft

Sinai

EGYPT
240,000 Troops
1,200 Tanks
500 Combat Aircraft

Dahab
Makna

UNITED NATIONS FORCE
978 Indians
795 Canadians
579 Yugoslavs
530 Swedes
430 Brazilians
61 Norwegians

Sharm el Sheikh

Red Sea

Strait of Tiran

Suez Canal

ALGERIA
60,000 Troops
400 Tanks
100 Combat Aircraft

CHRONOLOGY OF THE CRISIS

14 May: Egypt mobilizes

16 May: Egyptian forces move eastwards across Sinai

16 May: Egypt demands withdrawal of all United Nations forces

19 May: United Nations forces withdraw. Cairo radio announces: *'This is our chance Arabs, to deal Israel a mortal blow of annihilation, to blot out its entire presence in our holy land'*

22 May: Cairo radio announces: *'The Arab people is firmly resolved to wipe Israel off the map'.* Egypt declares the Strait of Tiran closed to all Israeli shipping

© Martin Gilbert

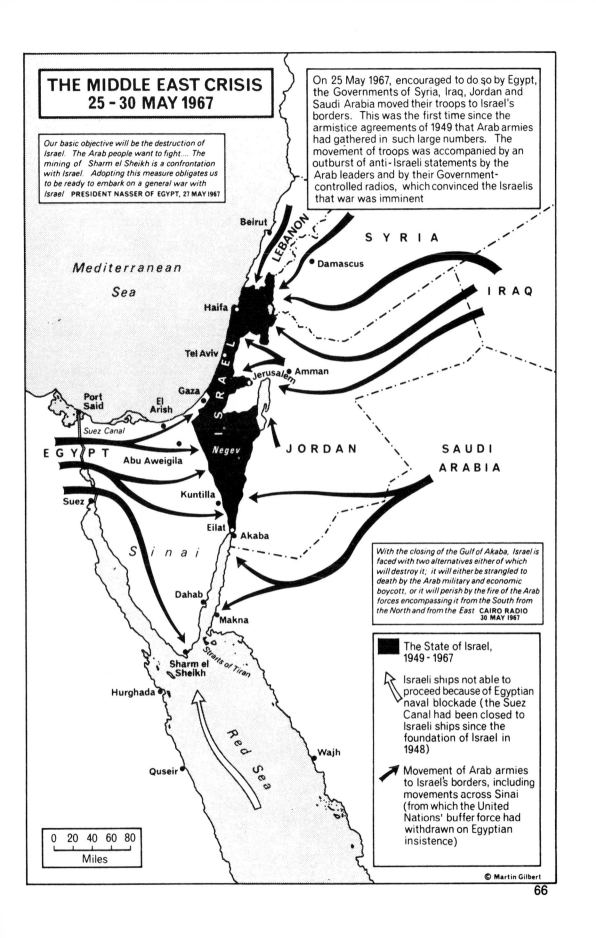

THE MIDDLE EAST CRISIS
25 - 30 MAY 1967

Our basic objective will be the destruction of Israel. The Arab people want to fight.... The mining of Sharm el Sheikh is a confrontation with Israel. Adopting this measure obligates us to be ready to embark on a general war with Israel PRESIDENT NASSER OF EGYPT, 27 MAY 1967

On 25 May 1967, encouraged to do so by Egypt, the Governments of Syria, Iraq, Jordan and Saudi Arabia moved their troops to Israel's borders. This was the first time since the armistice agreements of 1949 that Arab armies had gathered in such large numbers. The movement of troops was accompanied by an outburst of anti-Israeli statements by the Arab leaders and by their Government-controlled radios, which convinced the Israelis that war was imminent

Beirut

LEBANON

S Y R I A

• Damascus

I R A Q

Mediterranean Sea

Haifa

Tel Aviv

I S R A E L

Jerusalem • Amman

Gaza

Port Said

El Arish

Suez Canal

E G Y P T

Abu Aweigila

Negev

J O R D A N

S A U D I   A R A B I A

Suez

Kuntilla

Eilat

Akaba

S i n a i

Dahab

Straits of Tiran

• Makna

With the closing of the Gulf of Akaba, Israel is faced with two alternatives either of which will destroy it; it will either be strangled to death by the Arab military and economic boycott, or it will perish by the fire of the Arab forces encompassing it from the South from the North and from the East CAIRO RADIO 30 MAY 1967

Sharm el Sheikh

Hurghada •

Red Sea

Wajh

Quseir •

◼ The State of Israel, 1949 - 1967

⇧ Israeli ships not able to proceed because of Egyptian naval blockade (the Suez Canal had been closed to Israeli ships since the foundation of Israel in 1948)

➤ Movement of Arab armies to Israel's borders, including movements across Sinai (from which the United Nations' buffer force had withdrawn on Egyptian insistence)

0 20 40 60 80
Miles

© Martin Gilbert

66

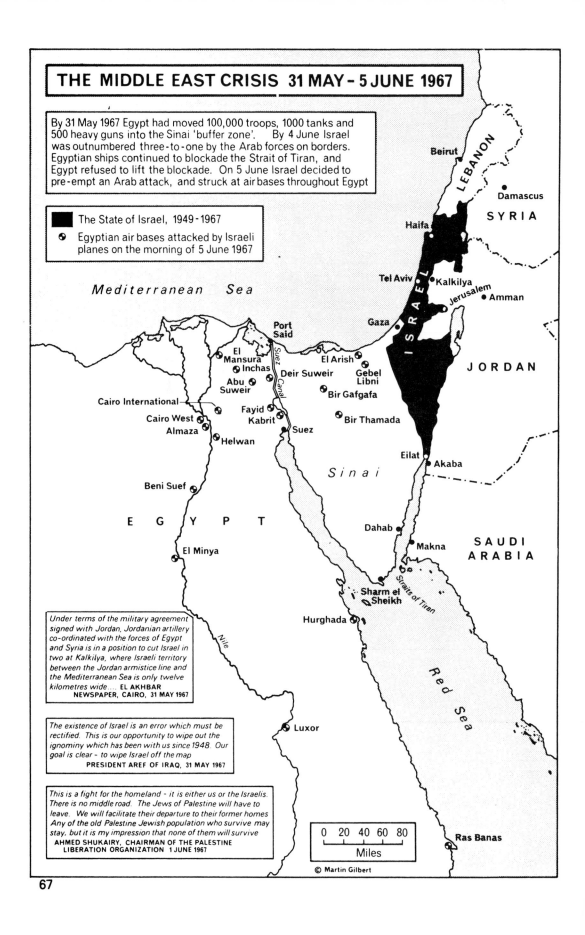

THE MIDDLE EAST CRISIS 31 MAY – 5 JUNE 1967

By 31 May 1967 Egypt had moved 100,000 troops, 1000 tanks and 500 heavy guns into the Sinai 'buffer zone'. By 4 June Israel was outnumbered three-to-one by the Arab forces on borders. Egyptian ships continued to blockade the Strait of Tiran, and Egypt refused to lift the blockade. On 5 June Israel decided to pre-empt an Arab attack, and struck at air bases throughout Egypt

■ The State of Israel, 1949-1967

⊙ Egyptian air bases attacked by Israeli planes on the morning of 5 June 1967

Mediterranean Sea

LEBANON

Beirut

Damascus

SYRIA

Haifa

Tel Aviv Kalkilya

ISRAEL

Jerusalem Amman

Gaza

JORDAN

Port Said

El Mansura

Inchas

El Arish

Deir Suweir

Gebel Libni

Abu Suweir

Bir Gafgafa

Cairo International

Fayid

Kabrit

Cairo West

Almaza

Suez

Bir Thamada

Helwan

Eilat Akaba

Beni Suef

Sinai

E G Y P T

Dahab

SAUDI ARABIA

Makna

El Minya

Nile

Under terms of the military agreement signed with Jordan, Jordanian artillery co-ordinated with the forces of Egypt and Syria is in a position to cut Israel in two at Kalkilya, where Israeli territory between the Jordan armistice line and the Mediterranean Sea is only twelve kilometres wide..... **EL AKHBAR NEWSPAPER, CAIRO, 31 MAY 1967**

Sharm el Sheikh

Straits of Tiran

Hurghada

Red Sea

The existence of Israel is an error which must be rectified. This is our opportunity to wipe out the ignominy which has been with us since 1948. Our goal is clear – to wipe Israel off the map **PRESIDENT AREF OF IRAQ, 31 MAY 1967**

Luxor

This is a fight for the homeland – it is either us or the Israelis. There is no middle road. The Jews of Palestine will have to leave. We will facilitate their departure to their former homes Any of the old Palestine Jewish population who survive may stay, but it is my impression that none of them will survive **AHMED SHUKAIRY, CHAIRMAN OF THE PALESTINE LIBERATION ORGANIZATION 1 JUNE 1967**

0 20 40 60 80

Miles

Ras Banas

© Martin Gilbert

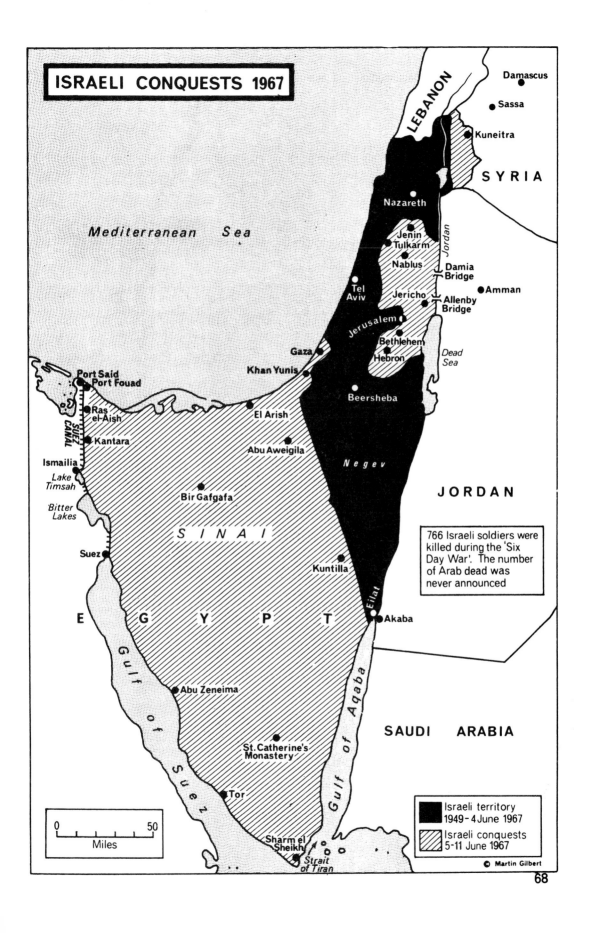

ISRAELI CONQUESTS 1967

Damascus
Sassa

LEBANON

Kuneitra

SYRIA

Mediterranean Sea

Nazareth

Jenin
Tulkarm
Nablus

Damia
Bridge

Jordan

Tel
Aviv

Jericho

Amman

Allenby
Bridge

Jerusalem

Bethlehem

Gaza

Hebron

Khan Yunis

Dead
Sea

Beersheba

El Arish

Abu Aweigila

Negev

JORDAN

Port Said
Port Fouad

Ras
el-Aish

Kantara

766 Israeli soldiers were
killed during the 'Six
Day War'. The number
of Arab dead was
never announced

Ismailia

Lake
Timsah

SUEZ CANAL

Bir Gafgafa

Bitter
Lakes

S I N A I

Suez

Kuntilla

E   G   Y   P   T

Eilat

Akaba

Abu Zeneima

Gulf of Suez

Gulf of Aqaba

SAUDI   ARABIA

St.Catherine's
Monastery

Tor

Sharm el
Sheikh

Israeli territory
1949–4 June 1967

Israeli conquests
5–11 June 1967

© Martin Gilbert

Strait
of Tiran

0   50

Miles

68

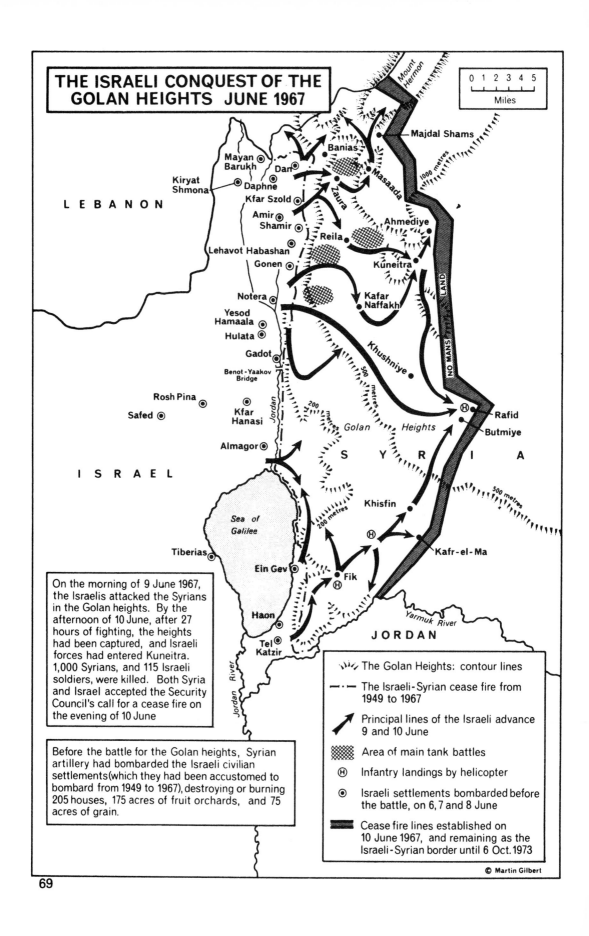

THE ISRAELI CONQUEST OF THE GOLAN HEIGHTS JUNE 1967

0 1 2 3 4 5
Miles

Mount Hermon

Majdal Shams

Mayan Barukh

Kiryat Shmona

Dan

Banias

Masaada

1000 metres

LEBANON

Daphne

Kfar Szold

Zaura

Amir

Shamir

Reila

Ahmediye

Lehavot Habashan

Kuneitra

Gonen

Notera

Kafar Naffakh

Yesod Hamaala

Hulata

NO MAN'S LAND

Gadot

Benot-Yaakov Bridge

Khushniye

500 metres

Rosh Pina

Kfar Hanasi

Jordan

Safed

200 metres

Golan   Heights

Rafid

Butmiye

Almagor

SYRIA

ISRAEL

Sea of Galilee

200 metres

500 metres

Khisfin

Tiberias

Ein Gev

Fik

Kafr-el-Ma

Haon

Yarmuk River

Tel Katzir

JORDAN

Jordan River

On the morning of 9 June 1967, the Israelis attacked the Syrians in the Golan heights. By the afternoon of 10 June, after 27 hours of fighting, the heights had been captured, and Israeli forces had entered Kuneitra. 1,000 Syrians, and 115 Israeli soldiers, were killed. Both Syria and Israel accepted the Security Council's call for a cease fire on the evening of 10 June

Before the battle for the Golan heights, Syrian artillery had bombarded the Israeli civilian settlements (which they had been accustomed to bombard from 1949 to 1967), destroying or burning 205 houses, 175 acres of fruit orchards, and 75 acres of grain.

	The Golan Heights: contour lines
—·—	The Israeli-Syrian cease fire from 1949 to 1967
↗	Principal lines of the Israeli advance 9 and 10 June
▨	Area of main tank battles
Ⓗ	Infantry landings by helicopter
◉	Israeli settlements bombarded before the battle, on 6, 7 and 8 June
	Cease fire lines established on 10 June 1967, and remaining as the Israeli-Syrian border until 6 Oct. 1973

© Martin Gilbert

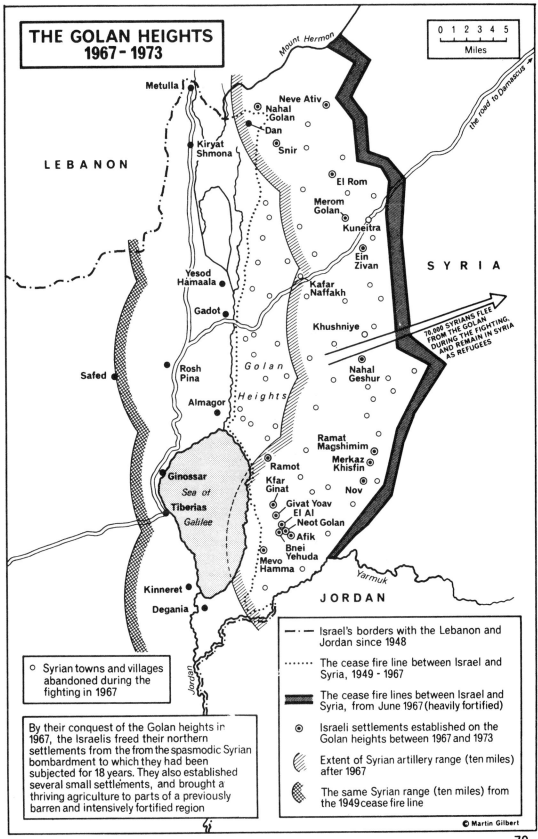

THE GOLAN HEIGHTS
1967-1973

0 1 2 3 4 5
Miles

Mount Hermon

the road to Damascus

Metulla

Neve Ativ
Nahal
Golan
Dan
Snir

Kiryat
Shmona

L E B A N O N

El Rom

Merom
Golan
Kuneitra

Ein
Zivan

S Y R I A

Yesod
Hamaala
Kafar
Naffakh

Gadot

Khushniye

70,000 SYRIANS FLEE
FROM THE GOLAN
DURING THE FIGHTING,
AND REMAIN IN SYRIA
AS REFUGEES

Golan

Nahal
Geshur

Safed

Rosh
Pina

Heights

Almagor

Ramat
Magshimim
Merkaz
Khisfin

Ramot
Kfar
Ginat

Nov

Ginossar

Givat Yoav
El Al
Neot Golan

Sea of

Tiberias

Galilee

Afik

Bnei
Yehuda
Mevo
Hamma

Yarmuk

Kinneret

J O R D A N

Degania

Jordan

○ Syrian towns and villages
abandoned during the
fighting in 1967

By their conquest of the Golan heights in
1967, the Israelis freed their northern
settlements from the from the spasmodic Syrian
bombardment to which they had been
subjected for 18 years. They also established
several small settlements, and brought a
thriving agriculture to parts of a previously
barren and intensively fortified region

–·–·– Israel's borders with the Lebanon and
Jordan since 1948

········· The cease fire line between Israel and
Syria, 1949 - 1967

▨▨▨ The cease fire lines between Israel and
Syria, from June 1967 (heavily fortified)

◉ Israeli settlements established on the
Golan heights between 1967 and 1973

▨ Extent of Syrian artillery range (ten miles)
after 1967

▨ The same Syrian range (ten miles) from
the 1949 cease fire line

© Martin Gilbert

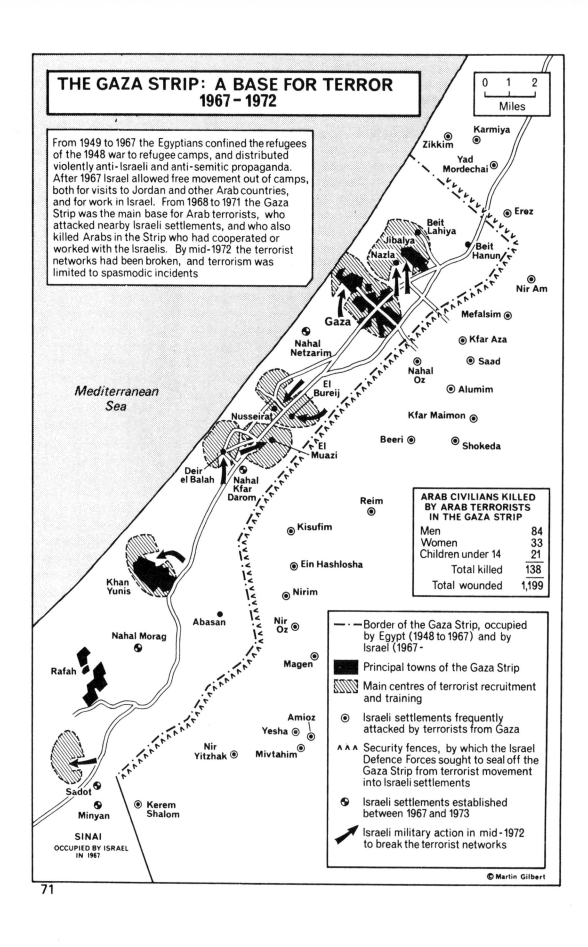

THE GAZA STRIP: A BASE FOR TERROR
1967 – 1972

From 1949 to 1967 the Egyptians confined the refugees
of the 1948 war to refugee camps, and distributed
violently anti-Israeli and anti-semitic propaganda.
After 1967 Israel allowed free movement out of camps,
both for visits to Jordan and other Arab countries,
and for work in Israel. From 1968 to 1971 the Gaza
Strip was the main base for Arab terrorists, who
attacked nearby Israeli settlements, and who also
killed Arabs in the Strip who had cooperated or
worked with the Israelis. By mid-1972 the terrorist
networks had been broken, and terrorism was
limited to spasmodic incidents

0 1 2
Miles

Karmiya
Zikkim
Yad
Mordechai
Erez
Beit
Lahiya
Jibalya
Nazla
Beit
Hanun
Nir Am
Gaza
Mefalsim
Kfar Aza
Nahal
Netzarim
Saad
Nahal
Oz
Alumim
El
Bureij
Kfar Maimon
Nusseirat
Beeri
Shokeda
El
Muazi

*Mediterranean
Sea*

Deir
el Balah
Nahal
Kfar
Darom
Reim

Kisufim

ARAB CIVILIANS KILLED
BY ARAB TERRORISTS
IN THE GAZA STRIP

Men	84
Women	33
Children under 14	21
Total killed	138
Total wounded	1,199

Ein Hashlosha

Khan
Yunis

Nirim

Abasan
Nir
Oz

Nahal Morag

Magen

Rafah

Amioz
Yesha
Nir
Yitzhak
Mivtahim

Sadot

Minyan
Kerem
Shalom

SINAI
OCCUPIED BY ISRAEL
IN 1967

— · —Border of the Gaza Strip, occupied
by Egypt (1948 to 1967) and by
Israel (1967 -

Principal towns of the Gaza Strip

Main centres of terrorist recruitment
and training

Israeli settlements frequently
attacked by terrorists from Gaza

Λ Λ Λ Security fences, by which the Israel
Defence Forces sought to seal off the
Gaza Strip from terrorist movement
into Israeli settlements

Israeli settlements established
between 1967 and 1973

Israeli military action in mid-1972
to break the terrorist networks

© Martin Gilbert

71

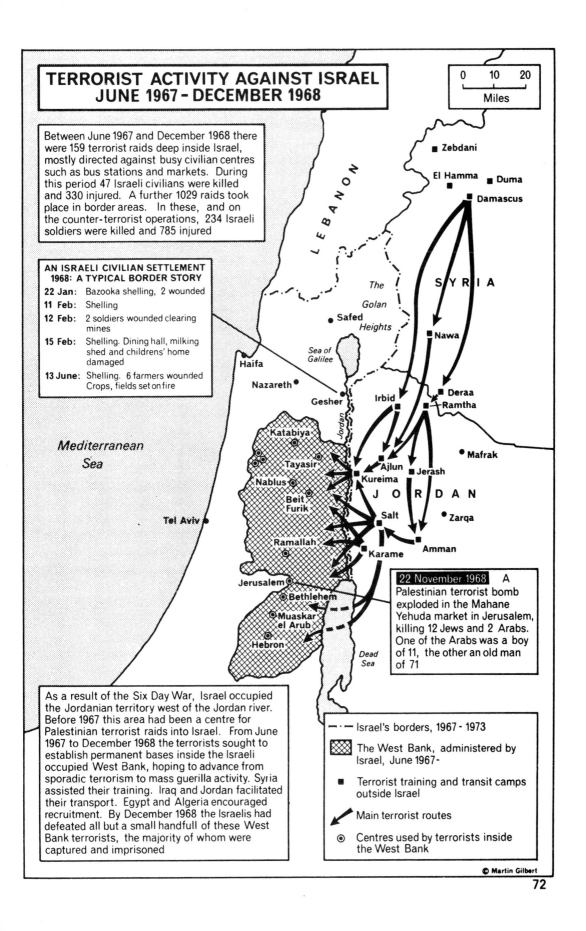

TERRORIST ACTIVITY AGAINST ISRAEL
JUNE 1967 - DECEMBER 1968

0   10   20
Miles

Between June 1967 and December 1968 there were 159 terrorist raids deep inside Israel, mostly directed against busy civilian centres such as bus stations and markets. During this period 47 Israeli civilians were killed and 330 injured. A further 1029 raids took place in border areas. In these, and on the counter-terrorist operations, 234 Israeli soldiers were killed and 785 injured

AN ISRAELI CIVILIAN SETTLEMENT 1968: A TYPICAL BORDER STORY

22 Jan: Bazooka shelling, 2 wounded

11 Feb: Shelling

12 Feb: 2 soldiers wounded clearing mines

15 Feb: Shelling. Dining hall, milking shed and childrens' home damaged

13 June: Shelling. 6 farmers wounded Crops, fields set on fire

L E B A N O N

S Y R I A

Zebdani

El Hamma ■   ■ Duma

■ Damascus

The
Golan

● Safed
Heights

Sea of
Galilee

■ Nawa

Haifa ●

Nazareth ●

Gesher

Irbid ■   ■ Deraa
         ← Ramtha

Mediterranean
Sea

Katabiya
◉

◉
◉
Tayasir ● ◉

Nablus ●◉

Beit
Furik ◉

Ramallah
◉

Jerusalem ◉

● Bethlehem

◉ Muaskar
el Arub

◉
Hebron

Ajlun ■   ■ Jerash
Kureima ■

J O R D A N

Salt ■

● Mafrak

● Zarqa

Amman ■

Karame

Tel Aviv ●

Dead
Sea

22 November 1968 A Palestinian terrorist bomb exploded in the Mahane Yehuda market in Jerusalem, killing 12 Jews and 2 Arabs. One of the Arabs was a boy of 11, the other an old man of 71

As a result of the Six Day War, Israel occupied the Jordanian territory west of the Jordan river. Before 1967 this area had been a centre for Palestinian terrorist raids into Israel. From June 1967 to December 1968 the terrorists sought to establish permanent bases inside the Israeli occupied West Bank, hoping to advance from sporadic terrorism to mass guerilla activity. Syria assisted their training. Iraq and Jordan facilitated their transport. Egypt and Algeria encouraged recruitment. By December 1968 the Israelis had defeated all but a small handfull of these West Bank terrorists, the majority of whom were captured and imprisoned

– · – Israel's borders, 1967 - 1973

▨ The West Bank, administered by Israel, June 1967-

■ Terrorist training and transit camps outside Israel

↙ Main terrorist routes

◉ Centres used by terrorists inside the West Bank

© Martin Gilbert

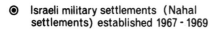

ISRAELI SECURITY MEASURES AND THE JORDAN VALLEY
1967 – 1970

⊙ Israeli military settlements (Nahal settlements) established 1967 - 1969

Ƹ Main Israeli border defences (forts, mine-fields, artillery units) established 1968-1970

Ⅱ Bridges open to Arab civilian trade and traffic in both directions, and across which Arabs suspected of terrorism were expelled

◖ Main areas of terrorist activity, June - December 1967

⬯ Main terrorist bases, for frequent raids across the Jordan, January - March 1968

⬯ Main terrorist bases (after the Karame raid) used for spasmodic raids across the Jordan, April 1968 - September 1970

▮ Towns and villages in which the Israeli military authorities blew up Arab houses between June 1967 and November 1969, as reprisals against terrorist activity. In all, more than 5,000 houses were destroyed, and more than five hundred Arabs expelled from the West Bank. By October 1973 more than a thousand Arabs had been expelled

–·–· The Israel - Jordan cease-fire line established in June 1967

······· The 'Green Line' cease-fire line between Israel and Jordan from 1949 to 1967

Armed struggle is the only way to liberate Palestine.... the liber-ation of Palestine is a national duty to repulse the Zionist, imperialist invasion from the great Arab homeland and to purge the Zionist presence from Palestine.... the partition of Palestine in 1947 and the establishment of Israel are fundamentally null and void.

THE PALESTINE NATIONAL COVENANT, adopted 17 JULY 1968 by the Palestine National Council

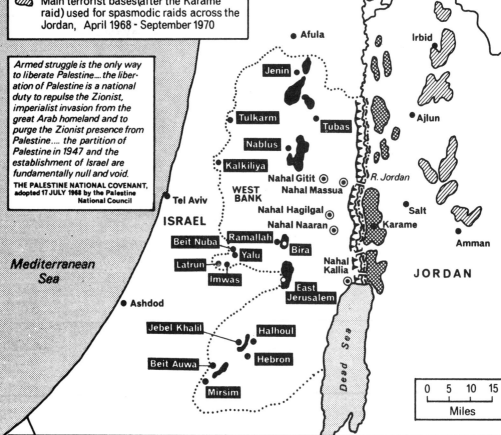

Afula · Irbid
Jenin
Tulkarm · Ajlun
Tubas
Nablus
Kalkiliya
Nahal Gitit ⊙ ⊙ R. Jordan
Nahal Massua
WEST BANK
Tel Aviv · Nahal Hagilgal ⊙
ISRAEL · Salt
Nahal Naaran ⊙ Karame
Beit Nuba · Ramallah
Yalu · Bira · Amman
Latrun · Nahal Kallia
Imwas · JORDAN
Mediterranean Sea · East Jerusalem
Ashdod ·
Jebel Khalil · Halhoul
Hebron
Beit Auwa
Mirsim · Dead Sea

```
0    5   10   15
|----|----|----|
      Miles
```

Between 1967 and 1970, Arab terrorists on the West Bank killed 12 Israelis, as well as over 50 Arabs whom they accused of 'collaborating' with Israel. Israeli forces were active in driving the terrorists towards the Jordan river, and on 21 March 1968 crossed the river in force to attack the terrorist base at Karame. Following this raid, the Israeli army established a fortified line along the Jordan, with a border fence and minefields, effectively sealing the border, and the terrorists themselves withdrew eastwards from the valley to the mountains. During 1970 terrorist acts on the West Bank stopped almost completely. They began again, on a smale scale, after October 1973

© Martin Gilbert

THE WEST BANK UNDER ISRAELI MILITARY ADMINISTRATION 1967–

The Israeli conquest of the West Bank in June 1967 brought 600,000 Arabs under Israeli military administration. The Israelis encouraged and financed economic development, and by the end of 1970 Arab unemployment had dropped from 12% to 3% By 1972 over 60,000 Arabs crossed the 'Green Line' every morning to work in Israel. At the same time, over 100,000 Arabs visited other Arab states for work, education and business. By 1973 over 14,500 West Bank Arabs were working in local administration in the West Bank

Sixteen Israeli settlements were founded in the West Bank between 1967 and 1973, with a total civilian population of 1,150. One of these settlements was near Hebron, where Jews had lived for more than two thousand years, before being driven out by the Arabs in 1929. Another group of settlements, the Etzion Bloc, was established on the site of settlements destroyed by the Arabs in 1949. At the same time, 44,000 Arabs who had fled from the West Bank in 1967, returned by 1972

During 1970 an 'Open Bridges' policy enabled West Bank Arabs to cross freely into Jordan. At the height of the tourist season in 1970, some 750 buses a day crossed from Jordan to Hebron, and by the end of August 50,000 Arabs had visited the West Bank from Jordan

ARABS VISITING THE WEST BANK FROM OTHER ARAB AREAS	
1968	16,000
1969	23,000
1970	52,000
1971	107,000
1972	150,000

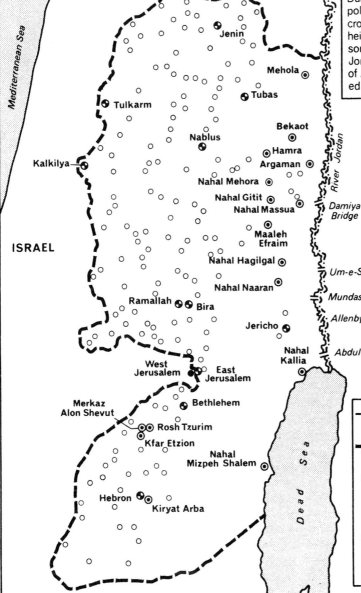

- –·– The border between Jordan and Israel, 1967
- ▬▬ The 'Green Line', Israel's border with Jordan, 1949–1967
- ◉ Israeli settlements established in the 'administered territories' between 1967 and 1973
- ✪ Principal Arab towns on the West Bank
- ○ Main Arab villages

© Martin Gilbert

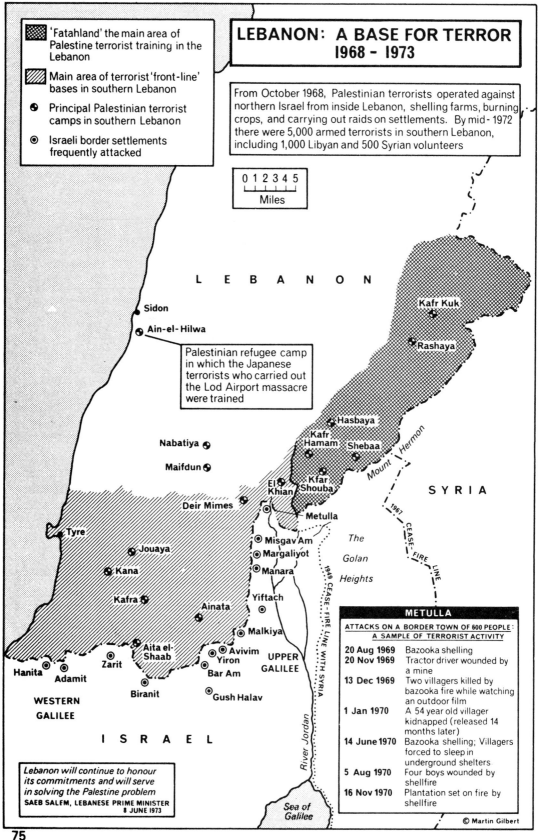

LEBANON: A BASE FOR TERROR 1968 - 1973

Legend

- 'Fatahland' the main area of Palestine terrorist training in the Lebanon
- Main area of terrorist 'front-line' bases in southern Lebanon
- Principal Palestinian terrorist camps in southern Lebanon
- Israeli border settlements frequently attacked

From October 1968, Palestinian terrorists operated against northern Israel from inside Lebanon, shelling farms, burning crops, and carrying out raids on settlements. By mid-1972 there were 5,000 armed terrorists in southern Lebanon, including 1,000 Libyan and 500 Syrian volunteers

0 1 2 3 4 5
Miles

LEBANON

Sidon

Ain-el-Hilwa

Palestinian refugee camp in which the Japanese terrorists who carried out the Lod Airport massacre were trained

Kafr Kuk

Rashaya

Nabatiya

Maifdun

Hasbaya

Kafr Hamam

Shebaa

El Khian

Kfar Shouba

SYRIA

Deir Mimes

Metulla

Mount Hermon

1967 CEASE-FIRE LINE

Tyre

Misgav Am

Margaliyot

Manara

The Golan Heights

Jouaya

Kana

Kafra

Ainata

Yiftach

Malkiya

1949 CEASE-FIRE LINE WITH SYRIA

Aita el-Shaab

Avivim

Yiron

Bar Am

UPPER GALILEE

Hanita

Adamit

Zarit

Biranit

Gush Halav

WESTERN GALILEE

ISRAEL

River Jordan

Sea of Galilee

Lebanon will continue to honour its commitments and will serve in solving the Palestine problem
SAEB SALEM, LEBANESE PRIME MINISTER 8 JUNE 1973

METULLA

ATTACKS ON A BORDER TOWN OF 600 PEOPLE: A SAMPLE OF TERRORIST ACTIVITY

20 Aug 1969	Bazooka shelling
20 Nov 1969	Tractor driver wounded by a mine
13 Dec 1969	Two villagers killed by bazooka fire while watching an outdoor film
1 Jan 1970	A 54 year old villager kidnapped (released 14 months later)
14 June 1970	Bazooka shelling; Villagers forced to sleep in underground shelters
5 Aug 1970	Four boys wounded by shellfire
16 Nov 1970	Plantation set on fire by shellfire

© Martin Gilbert

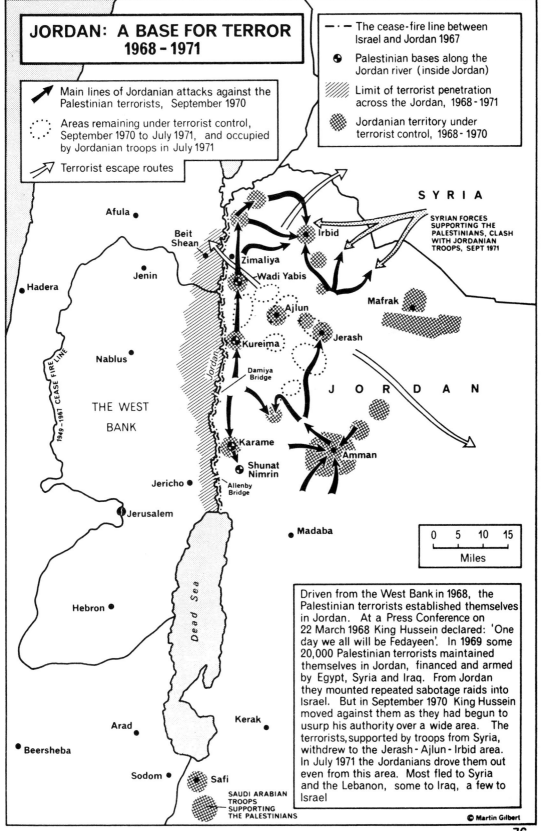

JORDAN: A BASE FOR TERROR 1968 – 1971

Main lines of Jordanian attacks against the Palestinian terrorists, September 1970

Areas remaining under terrorist control, September 1970 to July 1971, and occupied by Jordanian troops in July 1971

Terrorist escape routes

— · — The cease-fire line between Israel and Jordan 1967

✪ Palestinian bases along the Jordan river (inside Jordan)

▨ Limit of terrorist penetration across the Jordan, 1968-1971

▨ Jordanian territory under terrorist control, 1968-1970

S Y R I A

Afula

Beit Shean

Jenin

Hadera

Zimaliya

Wadi Yabis

Irbid

SYRIAN FORCES SUPPORTING THE PALESTINIANS, CLASH WITH JORDANIAN TROOPS, SEPT 1971

Mafrak

Ajlun

Jerash

Nablus

Kureima

Damiya Bridge

J O R D A N

1949 – 1967 CEASE FIRE LINE

THE WEST BANK

Jordan

Karame

Shunat Nimrin

Amman

Jericho

Allenby Bridge

Jerusalem

Madaba

Hebron

Dead Sea

0   5   10   15

Miles

Arad

Kerak

Beersheba

Sodom

Safi

SAUDI ARABIAN TROOPS SUPPORTING THE PALESTINIANS

Driven from the West Bank in 1968, the Palestinian terrorists established themselves in Jordan. At a Press Conference on 22 March 1968 King Hussein declared: 'One day we all will be Fedayeen'. In 1969 some 20,000 Palestinian terrorists maintained themselves in Jordan, financed and armed by Egypt, Syria and Iraq. From Jordan they mounted repeated sabotage raids into Israel. But in September 1970 King Hussein moved against them as they had begun to usurp his authority over a wide area. The terrorists, supported by troops from Syria, withdrew to the Jerash - Ajlun - Irbid area. In July 1971 the Jordanians drove them out even from this area. Most fled to Syria and the Lebanon, some to Iraq, a few to Israel

© Martin Gilbert

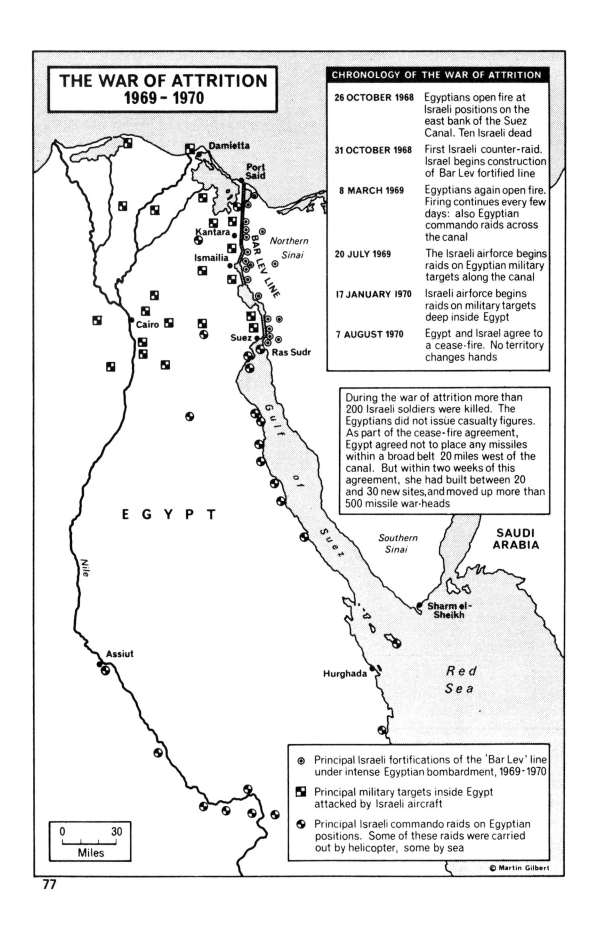

THE WAR OF ATTRITION
1969 - 1970

CHRONOLOGY OF THE WAR OF ATTRITION

26 OCTOBER 1968	Egyptians open fire at Israeli positions on the east bank of the Suez Canal. Ten Israeli dead
31 OCTOBER 1968	First Israeli counter-raid. Israel begins construction of Bar Lev fortified line
8 MARCH 1969	Egyptians again open fire. Firing continues every few days: also Egyptian commando raids across the canal
20 JULY 1969	The Israeli airforce begins raids on Egyptian military targets along the canal
17 JANUARY 1970	Israeli airforce begins raids on military targets deep inside Egypt
7 AUGUST 1970	Egypt and Israel agree to a cease-fire. No territory changes hands

During the war of attrition more than 200 Israeli soldiers were killed. The Egyptians did not issue casualty figures. As part of the cease-fire agreement, Egypt agreed not to place any missiles within a broad belt 20 miles west of the canal. But within two weeks of this agreement, she had built between 20 and 30 new sites, and moved up more than 500 missile war-heads

Damietta

Port Said

Kantara

BAR LEV LINE

Northern Sinai

Ismailia

Cairo

Suez

Ras Sudr

Gulf of Suez

E G Y P T

Nile

Southern Sinai

SAUDI ARABIA

Sharm el-Sheikh

Assiut

Hurghada

Red Sea

◉ Principal Israeli fortifications of the 'Bar Lev' line under intense Egyptian bombardment, 1969-1970

▣ Principal military targets inside Egypt attacked by Israeli aircraft

✪ Principal Israeli commando raids on Egyptian positions. Some of these raids were carried out by helicopter, some by sea

0   30
Miles

© Martin Gilbert

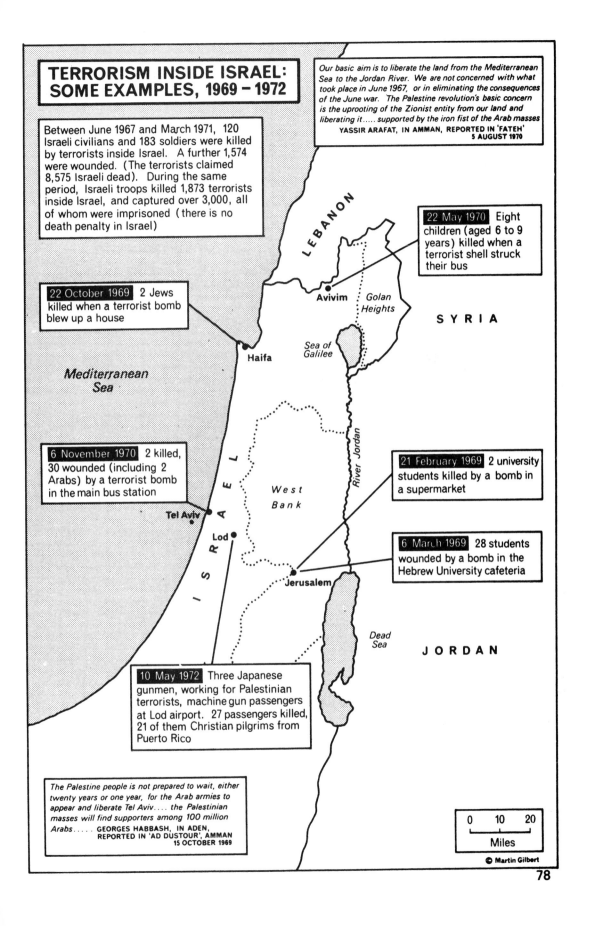

TERRORISM INSIDE ISRAEL: SOME EXAMPLES, 1969 – 1972

Our basic aim is to liberate the land from the Mediterranean Sea to the Jordan River. We are not concerned with what took place in June 1967, or in eliminating the consequences of the June war. The Palestine revolution's basic concern is the uprooting of the Zionist entity from our land and liberating it..... supported by the iron fist of the Arab masses
YASSIR ARAFAT, IN AMMAN, REPORTED IN 'FATEH'
5 AUGUST 1970

Between June 1967 and March 1971, 120 Israeli civilians and 183 soldiers were killed by terrorists inside Israel. A further 1,574 were wounded. (The terrorists claimed 8,575 Israeli dead). During the same period, Israeli troops killed 1,873 terrorists inside Israel, and captured over 3,000, all of whom were imprisoned (there is no death penalty in Israel)

22 May 1970 Eight children (aged 6 to 9 years) killed when a terrorist shell struck their bus

22 October 1969 2 Jews killed when a terrorist bomb blew up a house

LEBANON

Avivim

Golan Heights

S Y R I A

Sea of Galilee

Mediterranean Sea

Haifa

6 November 1970 2 killed, 30 wounded (including 2 Arabs) by a terrorist bomb in the main bus station

River Jordan

21 February 1969 2 university students killed by a bomb in a supermarket

W e s t B a n k

Tel Aviv

Lod

6 March 1969 28 students wounded by a bomb in the Hebrew University cafeteria

Jerusalem

I S R A E L

Dead Sea

J O R D A N

10 May 1972 Three Japanese gunmen, working for Palestinian terrorists, machine gun passengers at Lod airport. 27 passengers killed, 21 of them Christian pilgrims from Puerto Rico

The Palestine people is not prepared to wait, either twenty years or one year, for the Arab armies to appear and liberate Tel Aviv.... the Palestinian masses will find supporters among 100 million Arabs..... **GEORGES HABBASH, IN ADEN, REPORTED IN 'AD DUSTOUR', AMMAN 15 OCTOBER 1969**

0   10   20

Miles

© **Martin Gilbert**

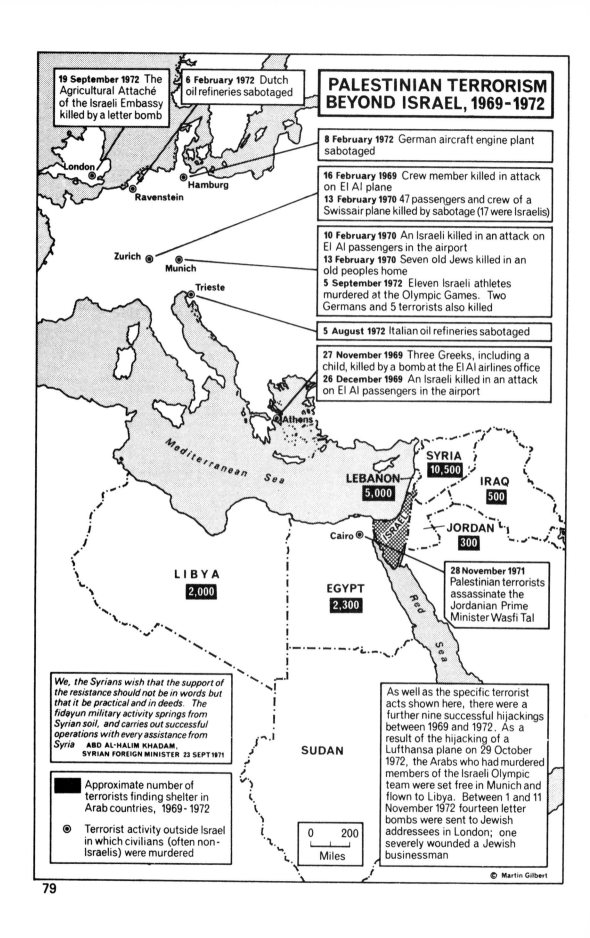

PALESTINIAN TERRORISM BEYOND ISRAEL, 1969-1972

19 September 1972 The Agricultural Attaché of the Israeli Embassy killed by a letter bomb

6 February 1972 Dutch oil refineries sabotaged

8 February 1972 German aircraft engine plant sabotaged

16 February 1969 Crew member killed in attack on El Al plane
13 February 1970 47 passengers and crew of a Swissair plane killed by sabotage (17 were Israelis)

10 February 1970 An Israeli killed in an attack on El Al passengers in the airport
13 February 1970 Seven old Jews killed in an old peoples home
5 September 1972 Eleven Israeli athletes murdered at the Olympic Games. Two Germans and 5 terrorists also killed

5 August 1972 Italian oil refineries sabotaged

27 November 1969 Three Greeks, including a child, killed by a bomb at the El Al airlines office
26 December 1969 An Israeli killed in an attack on El Al passengers in the airport

London

Ravenstein

Hamburg

Zurich

Munich

Trieste

Mediterranean Sea

Athens

SYRIA `10,500`

LEBANON `5,000`

IRAQ `500`

JORDAN `300`

Cairo

28 November 1971 Palestinian terrorists assassinate the Jordanian Prime Minister Wasfi Tal

LIBYA `2,000`

EGYPT `2,300`

ISRAEL

Red Sea

We, the Syrians wish that the support of the resistance should not be in words but that it be practical and in deeds. The fidayun military activity springs from Syrian soil, and carries out successful operations with every assistance from Syria **ABD AL-HALIM KHADAM, SYRIAN FOREIGN MINISTER 23 SEPT 1971**

SUDAN

As well as the specific terrorist acts shown here, there were a further nine successful hijackings between 1969 and 1972. As a result of the hijacking of a Lufthansa plane on 29 October 1972, the Arabs who had murdered members of the Israeli Olympic team were set free in Munich and flown to Libya. Between 1 and 11 November 1972 fourteen letter bombs were sent to Jewish addressees in London; one severely wounded a Jewish businessman

`■` Approximate number of terrorists finding shelter in Arab countries, 1969-1972

`◉` Terrorist activity outside Israel in which civilians (often non-Israelis) were murdered

0   200
Miles

© Martin Gilbert

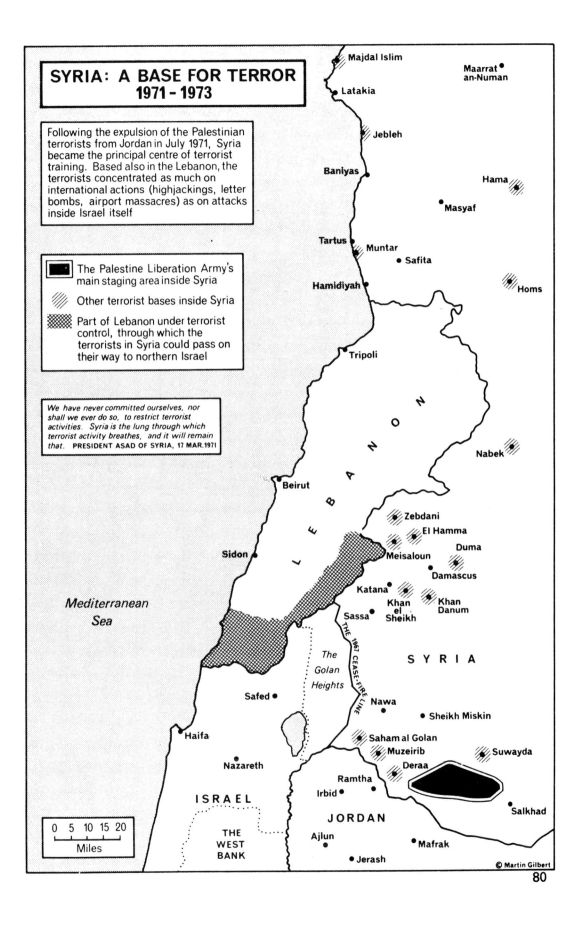

SYRIA: A BASE FOR TERROR 1971 – 1973

Following the expulsion of the Palestinian terrorists from Jordan in July 1971, Syria became the principal centre of terrorist training. Based also in the Lebanon, the terrorists concentrated as much on international actions (highjackings, letter bombs, airport massacres) as on attacks inside Israel itself

◼ The Palestine Liberation Army's main staging area inside Syria

▨ Other terrorist bases inside Syria

▦ Part of Lebanon under terrorist control, through which the terrorists in Syria could pass on their way to northern Israel

We have never committed ourselves, nor shall we ever do so, to restrict terrorist activities. Syria is the lung through which terrorist activity breathes, and it will remain that. **PRESIDENT ASAD OF SYRIA, 17 MAR.1971**

Majdal Islim
Maarrat an-Numan
Latakia
Jebleh
Baniyas
Hama
Masyaf
Tartus
Muntar
Safita
Hamidiyah
Homs
Tripoli
Nabek

LEBANON

Beirut

Zebdani
El Hamma
Duma
Meisaloun
Damascus
Sidon
Katana
Khan el Sheikh
Khan Danum
Sassa

Mediterranean Sea

SYRIA

The Golan Heights
Safed
Nawa
Sheikh Miskin
Haifa
Saham al Golan
Muzeirib
Suwayda
Deraa
Nazareth
Ramtha
Irbid
Salkhad

ISRAEL

JORDAN

THE WEST BANK
Ajlun
Mafrak
Jerash

0 5 10 15 20
Miles

THE 1967 CEASE-FIRE LINE

© Martin Gilbert

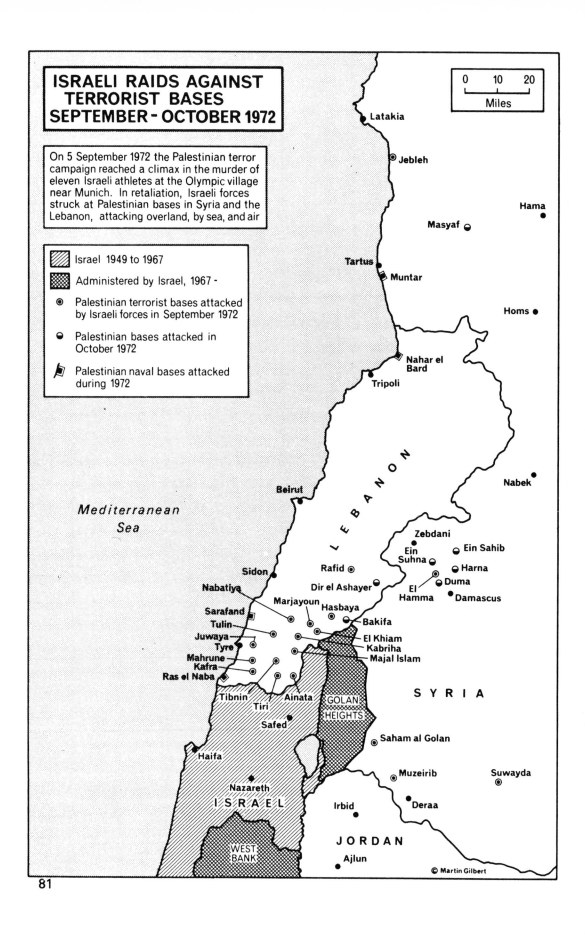

ISRAELI RAIDS AGAINST TERRORIST BASES SEPTEMBER – OCTOBER 1972

On 5 September 1972 the Palestinian terror campaign reached a climax in the murder of eleven Israeli athletes at the Olympic village near Munich. In retaliation, Israeli forces struck at Palestinian bases in Syria and the Lebanon, attacking overland, by sea, and air

Israel 1949 to 1967

Administered by Israel, 1967 -

◉ Palestinian terrorist bases attacked by Israeli forces in September 1972

◖ Palestinian bases attacked in October 1972

⚑ Palestinian naval bases attacked during 1972

0   10   20
Miles

Latakia

Jebleh

Hama

Masyaf

Tartus

Muntar

Homs

Nahar el Bard

Tripoli

Nabek

Beirut

Mediterranean Sea

L E B A N O N

Sidon

Zebdani

Ein Sahib

Rafid

Ein Suhna

Harna

Dir el Ashayer

Duma

El Hamma

Damascus

Nabatiya

Marjayoun

Hasbaya

Sarafand

Bakifa

Tulin

El Khiam

Juwaya

Kabriha

Tyre

Majal Islam

Mahrune

Kafra

Ras el Naba

S Y R I A

Tibnin

Ainata

Tiri

GOLAN HEIGHTS

Safed

Saham al Golan

Haifa

Muzeirib

Suwayda

Nazareth

I S R A E L

Irbid

Deraa

WEST BANK

J O R D A N

Ajlun

© Martin Gilbert

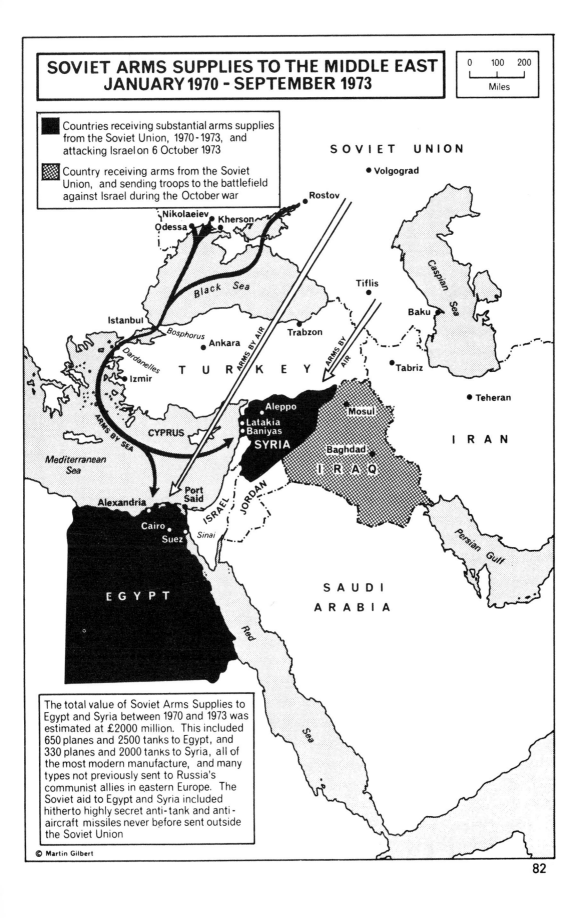

SOVIET ARMS SUPPLIES TO THE MIDDLE EAST
JANUARY 1970 - SEPTEMBER 1973

0   100   200
Miles

■ Countries receiving substantial arms supplies from the Soviet Union, 1970-1973, and attacking Israel on 6 October 1973

▨ Country receiving arms from the Soviet Union, and sending troops to the battlefield against Israel during the October war

SOVIET UNION

● Volgograd

● Rostov

Nikolaeiev
Odessa ● Kherson

Black Sea

Tiflis

Caspian Sea

Baku

Istanbul

Bosphorus ● Ankara   Trabzon

ARMS BY AIR

ARMS BY AIR

Tabriz

Dardanelles

Izmir

T U R K E Y

● Teheran

CYPRUS

ARMS BY SEA

Aleppo

Latakia
Baniyas

SYRIA

Mosul

I R A N

Mediterranean Sea

Baghdad

I R A Q

Alexandria

Port Said

ISRAEL

JORDAN

Persian Gulf

Cairo ●
Suez ●

Sinai

E G Y P T

S A U D I
A R A B I A

Red

Sea

The total value of Soviet Arms Supplies to Egypt and Syria between 1970 and 1973 was estimated at £2000 million. This included 650 planes and 2500 tanks to Egypt, and 330 planes and 2000 tanks to Syria, all of the most modern manufacture, and many types not previously sent to Russia's communist allies in eastern Europe. The Soviet aid to Egypt and Syria included hitherto highly secret anti-tank and anti-aircraft missiles never before sent outside the Soviet Union

© Martin Gilbert

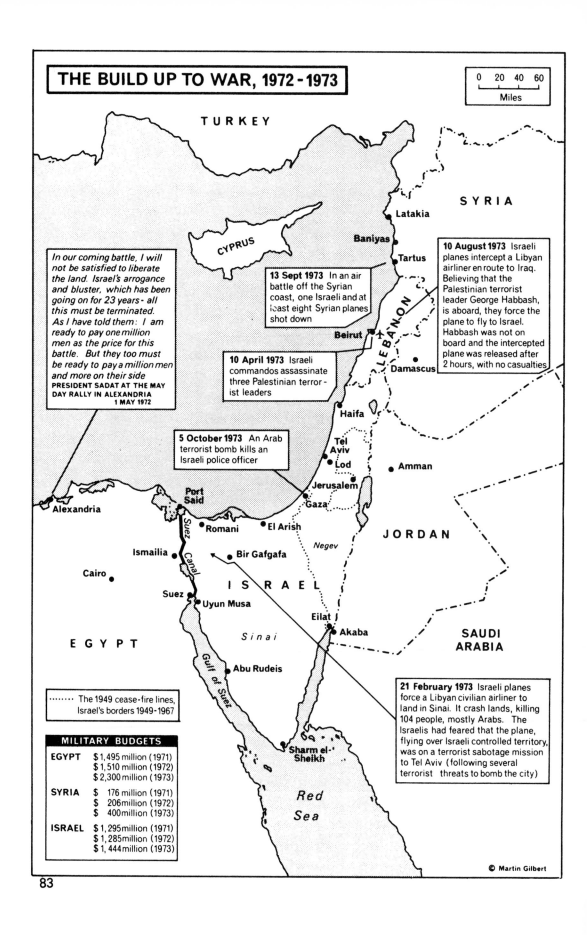

THE BUILD UP TO WAR, 1972-1973

0 20 40 60
Miles

TURKEY

CYPRUS

SYRIA

Latakia

Baniyas

Tartus

10 August 1973 Israeli planes intercept a Libyan airliner en route to Iraq. Believing that the Palestinian terrorist leader George Habbash, is aboard, they force the plane to fly to Israel. Habbash was not on board and the intercepted plane was released after 2 hours, with no casualties

In our coming battle, I will not be satisfied to liberate the land. Israel's arrogance and bluster, which has been going on for 23 years - all this must be terminated. As I have told them: I am ready to pay one million men as the price for this battle. But they too must be ready to pay a million men and more on their side **PRESIDENT SADAT AT THE MAY DAY RALLY IN ALEXANDRIA 1 MAY 1972**

13 Sept 1973 In an air battle off the Syrian coast, one Israeli and at least eight Syrian planes shot down

Beirut

LEBANON

Damascus

10 April 1973 Israeli commandos assassinate three Palestinian terrorist leaders

Haifa

Tel Aviv

Lod

Amman

5 October 1973 An Arab terrorist bomb kills an Israeli police officer

Jerusalem

Gaza

Port Said

Alexandria

Suez Canal

Romani

El Arish

Negev

JORDAN

Ismailia

Bir Gafgafa

Cairo

I S R A E L

Suez

Uyun Musa

Sinai

Eilat

Akaba

SAUDI ARABIA

E G Y P T

Gulf of Suez

Abu Rudeis

21 February 1973 Israeli planes force a Libyan civilian airliner to land in Sinai. It crash lands, killing 104 people, mostly Arabs. The Israelis had feared that the plane, flying over Israeli controlled territory, was on a terrorist sabotage mission to Tel Aviv (following several terrorist threats to bomb the city)

Sharm el-Sheikh

Red Sea

········ The 1949 cease-fire lines, Israel's borders 1949-1967

MILITARY BUDGETS		
EGYPT	$ 1,495 million (1971)	
	$ 1,510 million (1972)	
	$ 2,300 million (1973)	
SYRIA	$ 176 million (1971)	
	$ 206 million (1972)	
	$ 400 million (1973)	
ISRAEL	$ 1,295 million (1971)	
	$ 1,285 million (1972)	
	$ 1,444 million (1973)	

© Martin Gilbert

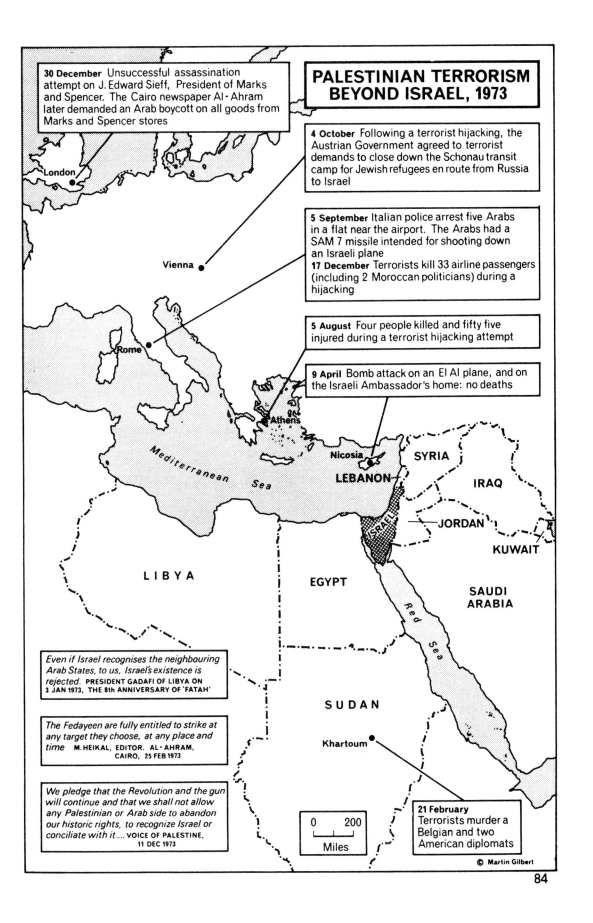

PALESTINIAN TERRORISM BEYOND ISRAEL, 1973

30 December Unsuccessful assassination attempt on J. Edward Sieff, President of Marks and Spencer. The Cairo newspaper Al-Ahram later demanded an Arab boycott on all goods from Marks and Spencer stores

4 October Following a terrorist hijacking, the Austrian Government agreed to terrorist demands to close down the Schonau transit camp for Jewish refugees en route from Russia to Israel

5 September Italian police arrest five Arabs in a flat near the airport. The Arabs had a SAM 7 missile intended for shooting down an Israeli plane
17 December Terrorists kill 33 airline passengers (including 2 Moroccan politicians) during a hijacking

5 August Four people killed and fifty five injured during a terrorist hijacking attempt

9 April Bomb attack on an El Al plane, and on the Israeli Ambassador's home: no deaths

London

Vienna

Rome

Athens

Mediterranean Sea

Nicosia

SYRIA

LEBANON

IRAQ

ISRAEL

JORDAN

KUWAIT

LIBYA

EGYPT

SAUDI ARABIA

Red Sea

Even if Israel recognises the neighbouring Arab States, to us, Israel's existence is rejected. **PRESIDENT GADAFI OF LIBYA ON 3 JAN 1973, THE 8th ANNIVERSARY OF 'FATAH'**

The Fedayeen are fully entitled to strike at any target they choose, at any place and time **M.HEIKAL, EDITOR. AL-AHRAM, CAIRO, 25 FEB 1973**

We pledge that the Revolution and the gun will continue and that we shall not allow any Palestinian or Arab side to abandon our historic rights, to recognize Israel or conciliate with it.... **VOICE OF PALESTINE, 11 DEC 1973**

SUDAN

Khartoum

0 200
Miles

21 February Terrorists murder a Belgian and two American diplomats

© Martin Gilbert

84

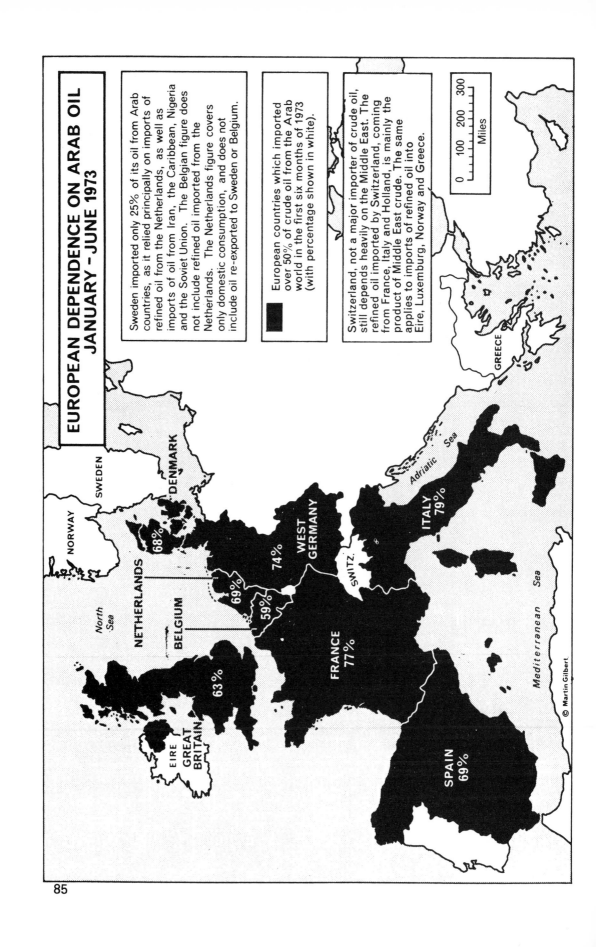

EUROPEAN DEPENDENCE ON ARAB OIL
JANUARY – JUNE 1973

Sweden imported only 25% of its oil from Arab countries, as it relied principally on imports of refined oil from the Netherlands, as well as imports of oil from Iran, the Caribbean, Nigeria and the Soviet Union. The Belgian figure does not include refined oil imported from the Netherlands. The Netherlands figure covers only domestic consumption, and does not include oil re-exported to Sweden or Belgium.

European countries which imported over 50% of crude oil from the Arab world in the first six months of 1973 (with percentage shown in white).

Switzerland, not a major importer of crude oil, still depends heavily on the Middle East. The refined oil imported by Switzerland, coming from France, Italy and Holland, is mainly the product of Middle East crude. The same applies to imports of refined oil into Eire, Luxemburg, Norway and Greece.

0 100 200 300
Miles

NORWAY

SWEDEN

DENMARK 68%

North Sea

NETHERLANDS 69%

BELGIUM 59%

WEST GERMANY 74%

SWITZ.

EIRE

GREAT BRITAIN 63%

FRANCE 77%

SPAIN 69%

ITALY 79%

Adriatic Sea

Mediterranean Sea

GREECE

© Martin Gilbert.

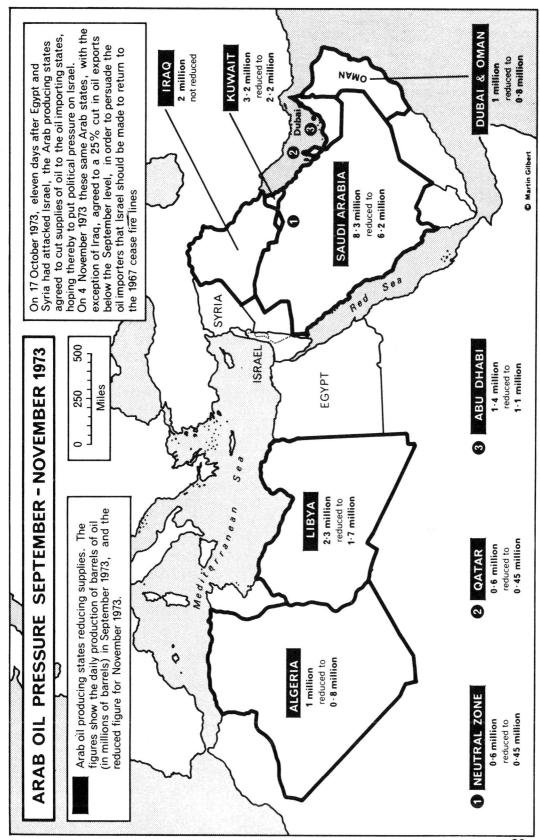

ARAB OIL PRESSURE SEPTEMBER – NOVEMBER 1973

Arab oil producing states reducing supplies. The figures show the daily production of barrels of oil (in millions of barrels) in September 1973, and the reduced figure for November 1973.

On 17 October 1973, eleven days after Egypt and Syria had attacked Israel, the Arab producing states agreed to cut supplies of oil to the oil importing states, hoping thereby to put political pressure on Israel.
On 4 November 1973 these same Arab states, with the exception of Iraq, agreed to a 25% cut in oil exports below the September level, in order to persuade the oil importers that Israel should be made to return to the 1967 cease fire lines

Miles

0   250   500

IRAQ
2 million
not reduced

KUWAIT
3·2 million
reduced to
2·2 million

DUBAI & OMAN
1 million
reduced to
0·8 million

SAUDI ARABIA
8·3 million
reduced to
6·2 million

OMAN

Dubai

Red Sea

SYRIA

ISRAEL

EGYPT

Mediterranean Sea

ABU DHABI
1·4 million
reduced to
1·1 million

LIBYA
2·3 million
reduced to
1·7 million

QATAR
0·6 million
reduced to
0·45 million

ALGERIA
1 million
reduced to
0·8 million

NEUTRAL ZONE
0·6 million
reduced to
0·45 million

© Martin Gilbert

86

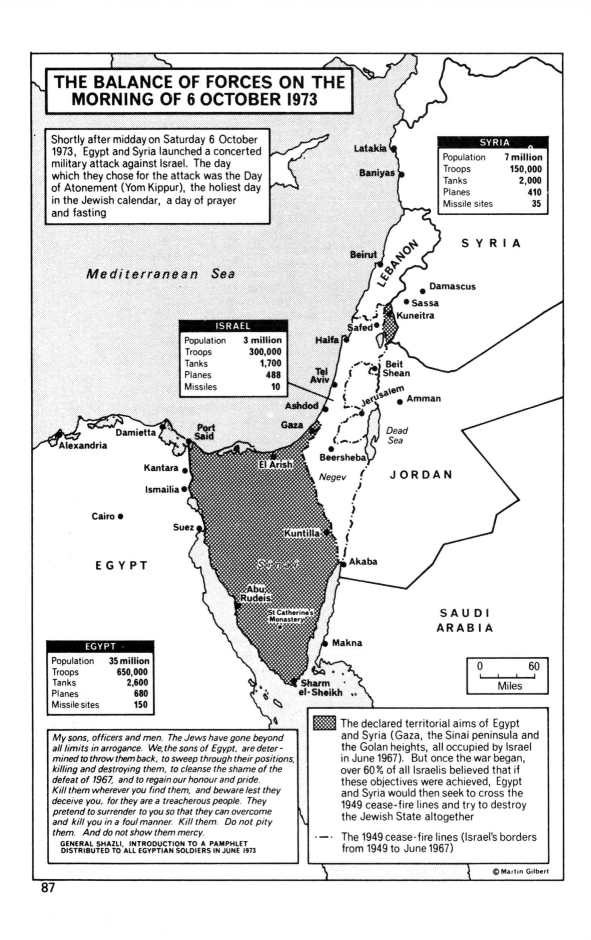

THE BALANCE OF FORCES ON THE MORNING OF 6 OCTOBER 1973

Shortly after midday on Saturday 6 October 1973, Egypt and Syria launched a concerted military attack against Israel. The day which they chose for the attack was the Day of Atonement (Yom Kippur), the holiest day in the Jewish calendar, a day of prayer and fasting

SYRIA	
Population	7 million
Troops	150,000
Tanks	2,000
Planes	410
Missile sites	35

ISRAEL	
Population	3 million
Troops	300,000
Tanks	1,700
Planes	488
Missiles	10

EGYPT	
Population	35 million
Troops	650,000
Tanks	2,600
Planes	680
Missile sites	150

Mediterranean Sea

Latakia
Baniyas
Beirut
LEBANON
SYRIA
Damascus
Sassa
Kuneitra
Safed
Haifa
Beit Shean
Tel Aviv
Jerusalem
Amman
Ashdod
Gaza
Dead Sea
Port Said
Damietta
Alexandria
Kantara
Ismailia
Beersheba
Negev
JORDAN
Cairo
Suez
Kuntilla
Akaba
EGYPT
Sinai
Abu Rudeis
St Catherine's Monastery
SAUDI ARABIA
Makna
Sharm el-Sheikh

0   60
Miles

My sons, officers and men. The Jews have gone beyond all limits in arrogance. We, the sons of Egypt, are determined to throw them back, to sweep through their positions, killing and destroying them, to cleanse the shame of the defeat of 1967, and to regain our honour and pride. Kill them wherever you find them, and beware lest they deceive you, for they are a treacherous people. They pretend to surrender to you so that they can overcome and kill you in a foul manner. Kill them. Do not pity them. And do not show them mercy.
GENERAL SHAZLI, INTRODUCTION TO A PAMPHLET DISTRIBUTED TO ALL EGYPTIAN SOLDIERS IN JUNE 1973

The declared territorial aims of Egypt and Syria (Gaza, the Sinai peninsula and the Golan heights, all occupied by Israel in June 1967). But once the war began, over 60% of all Israelis believed that if these objectives were achieved, Egypt and Syria would then seek to cross the 1949 cease-fire lines and try to destroy the Jewish State altogether

- - - The 1949 cease-fire lines (Israel's borders from 1949 to June 1967)

© Martin Gilbert

87

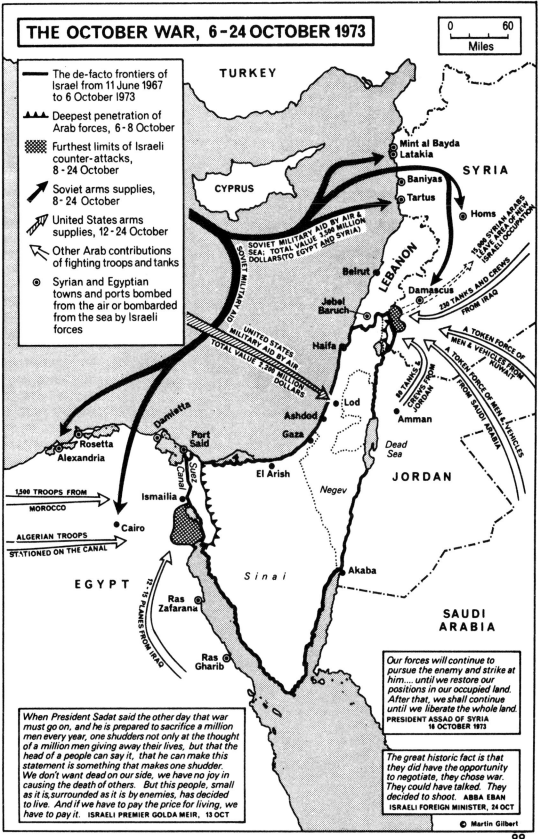

THE OCTOBER WAR, 6-24 OCTOBER 1973

0   60

Miles

—— The de-facto frontiers of Israel from 11 June 1967 to 6 October 1973

▲▲▲ Deepest penetration of Arab forces, 6-8 October

▓▓▓ Furthest limits of Israeli counter-attacks, 8-24 October

➤ Soviet arms supplies, 8-24 October

⇗ United States arms supplies, 12-24 October

⇦ Other Arab contributions of fighting troops and tanks

◉ Syrian and Egyptian towns and ports bombed from the air or bombarded from the sea by Israeli forces

TURKEY

CYPRUS

SYRIA

Mint al Bayda
Latakia
Baniyas
Tartus
Homs

15,000 SYRIAN ARABS LEAVE AREA OF NEW ISRAELI OCCUPATION

SOVIET MILITARY AID BY AIR & SEA; TOTAL VALUE 3,500 MILLION DOLLARS (TO EGYPT AND SYRIA)

SOVIET MILITARY AID

Beirut

LEBANON

Damascus

230 TANKS AND CREWS FROM IRAQ

Jebel Baruch

A TOKEN FORCE OF MEN & VEHICLES FROM KUWAIT

UNITED STATES MILITARY AID BY AIR TOTAL VALUE 2,200 MILLION DOLLARS

Haifa

80 TANKS & CREWS FROM JORDAN

A TOKEN FORCE OF MEN & VEHICLES FROM SAUDI ARABIA

Lod

Amman

Ashdod
Gaza

Dead Sea

JORDAN

Damietta
Port Said

Rosetta
Alexandria

El Arish

Negev

Suez Canal

1,500 TROOPS FROM MOROCCO

Ismailia

ALGERIAN TROOPS STATIONED ON THE CANAL

Cairo

Akaba

E G Y P T

S i n a i

12-15 PLANES FROM IRAQ

Ras Zafarana

SAUDI ARABIA

Ras Gharib

Our forces will continue to pursue the enemy and strike at him..... until we restore our positions in our occupied land. After that, we shall continue until we liberate the whole land.
PRESIDENT ASSAD OF SYRIA 16 OCTOBER 1973

When President Sadat said the other day that war must go on, and he is prepared to sacrifice a million men every year, one shudders not only at the thought of a million men giving away their lives, but that the head of a people can say it, that he can make this statement is something that makes one shudder. We don't want dead on our side, we have no joy in causing the death of others. But this people, small as it is, surrounded as it is by enemies, has decided to live. And if we have to pay the price for living, we have to pay it. **ISRAELI PREMIER GOLDA MEIR, 13 OCT**

The great historic fact is that they did have the opportunity to negotiate, they chose war. They could have talked. They decided to shoot. **ABBA EBAN ISRAELI FOREIGN MINISTER, 24 OCT**

© Martin Gilbert

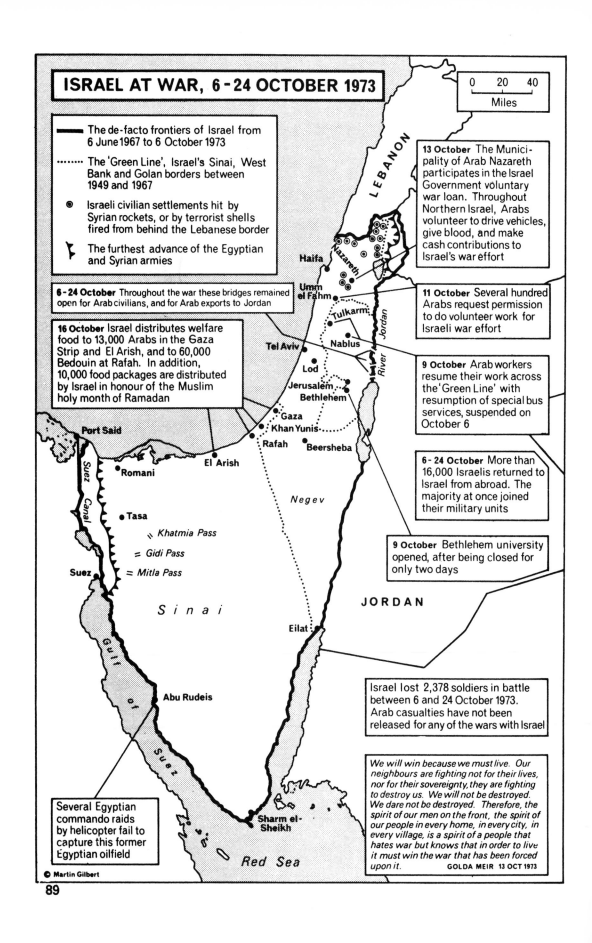

ISRAEL AT WAR, 6-24 OCTOBER 1973

0   20   40
Miles

—— The de-facto frontiers of Israel from 6 June 1967 to 6 October 1973

········ The 'Green Line', Israel's Sinai, West Bank and Golan borders between 1949 and 1967

◉ Israeli civilian settlements hit by Syrian rockets, or by terrorist shells fired from behind the Lebanese border

ʓ The furthest advance of the Egyptian and Syrian armies

6-24 October Throughout the war these bridges remained open for Arab civilians, and for Arab exports to Jordan

16 October Israel distributes welfare food to 13,000 Arabs in the Gaza Strip and El Arish, and to 60,000 Bedouin at Rafah. In addition, 10,000 food packages are distributed by Israel in honour of the Muslim holy month of Ramadan

13 October The Municipality of Arab Nazareth participates in the Israel Government voluntary war loan. Throughout Northern Israel, Arabs volunteer to drive vehicles, give blood, and make cash contributions to Israel's war effort

11 October Several hundred Arabs request permission to do volunteer work for Israeli war effort

9 October Arab workers resume their work across the 'Green Line' with resumption of special bus services, suspended on October 6

6-24 October More than 16,000 Israelis returned to Israel from abroad. The majority at once joined their military units

9 October Bethlehem university opened, after being closed for only two days

Israel lost 2,378 soldiers in battle between 6 and 24 October 1973. Arab casualties have not been released for any of the wars with Israel

We will win because we must live. Our neighbours are fighting not for their lives, nor for their sovereignty, they are fighting to destroy us. We will not be destroyed. We dare not be destroyed. Therefore, the spirit of our men on the front, the spirit of our people in every home, in every city, in every village, is a spirit of a people that hates war but knows that in order to live it must win the war that has been forced upon it. **GOLDA MEIR 13 OCT 1973**

Several Egyptian commando raids by helicopter fail to capture this former Egyptian oilfield

LEBANON

Haifa

Nazareth

Umm el Fahm

Tulkarm

Tel Aviv

Nablus

River Jordan

Lod

Jerusalem

Bethlehem

Gaza

Khan Yunis

Rafah

Beersheba

Port Said

Suez Canal

Romani

El Arish

Tasa

Khatmia Pass

Gidi Pass

Mitla Pass

Suez

Negev

S i n a i

JORDAN

Eilat

Gulf of Suez

Abu Rudeis

Sharm el-Sheikh

Red Sea

© Martin Gilbert

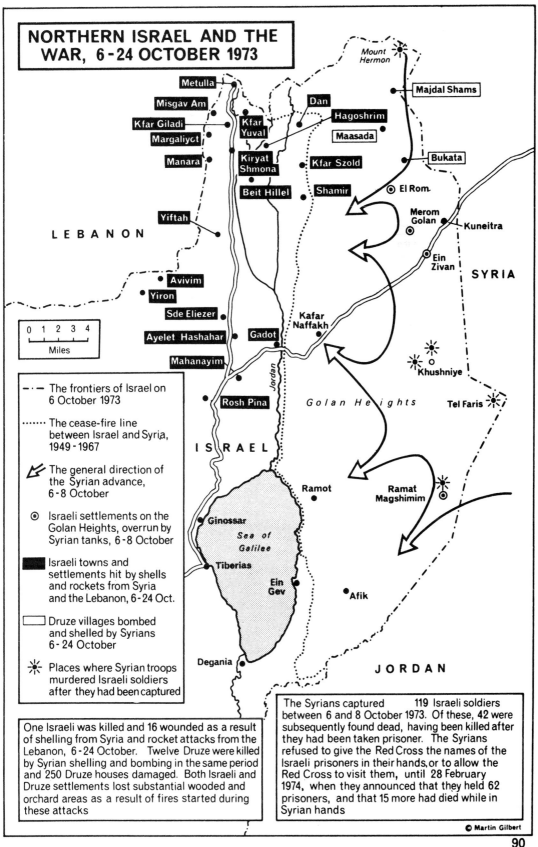

NORTHERN ISRAEL AND THE WAR, 6-24 OCTOBER 1973

Mount Hermon

Metulla

Majdal Shams

Misgav Am

Dan

Kfar Giladi

Kfar Yuval

Hagoshrim

Margaliyot

Maasada

Bukata

Manara

Kiryat Shmona

Kfar Szold

Beit Hillel

Shamir

⊙ El Rom.

Yiftah

Merom Golan

Kuneitra

L E B A N O N

⊙ Merom Golan

Ein Zivan ⊙

S Y R I A

Avivim

Yiron

Sde Eliezer

Kafar Naffakh

Ayelet Hashahar

Gadot

☀ Khushniye

Mahanayim

Jordan

Golan Heights

Tel Faris ☀

Rosh Pina

- · — The frontiers of Israel on 6 October 1973

······ The cease-fire line between Israel and Syria, 1949-1967

↙ The general direction of the Syrian advance, 6-8 October

⊙ Israeli settlements on the Golan Heights, overrun by Syrian tanks, 6-8 October

■ Israeli towns and settlements hit by shells and rockets from Syria and the Lebanon, 6-24 Oct.

▢ Druze villages bombed and shelled by Syrians 6-24 October

☀ Places where Syrian troops murdered Israeli soldiers after they had been captured

I S R A E L

Ramot

Ramat Magshimim ☀⊙

Ginossar

Sea of Galilee

Tiberias

Ein Gev

Afik

Degania

J O R D A N

Scale: 0 1 2 3 4 Miles

One Israeli was killed and 16 wounded as a result of shelling from Syria and rocket attacks from the Lebanon, 6-24 October. Twelve Druze were killed by Syrian shelling and bombing in the same period and 250 Druze houses damaged. Both Israeli and Druze settlements lost substantial wooded and orchard areas as a result of fires started during these attacks

The Syrians captured 119 Israeli soldiers between 6 and 8 October 1973. Of these, 42 were subsequently found dead, having been killed after they had been taken prisoner. The Syrians refused to give the Red Cross the names of the Israeli prisoners in their hands, or to allow the Red Cross to visit them, until 28 February 1974, when they announced that they held 62 prisoners, and that 15 more had died while in Syrian hands

© Martin Gilbert

90

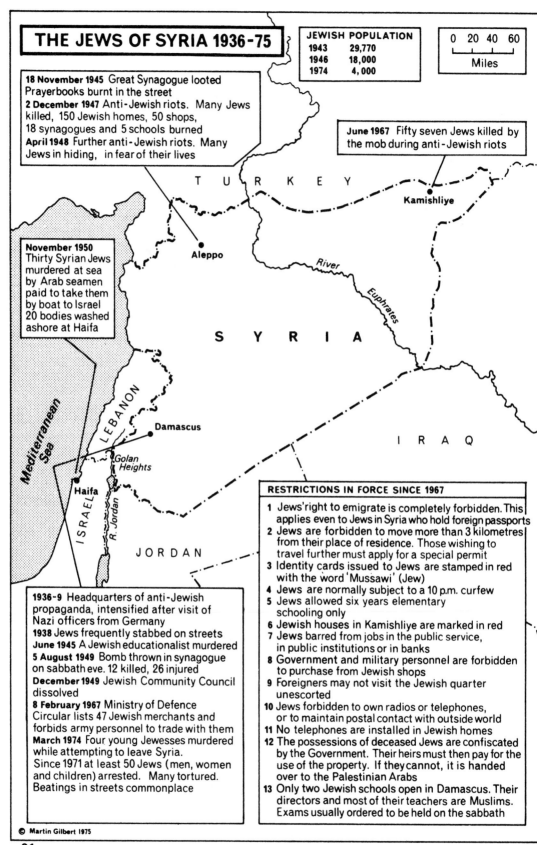

THE JEWS OF SYRIA 1936-75

JEWISH POPULATION
1943   29,770
1946   18,000
1974   4,000

0   20   40   60
Miles

18 November 1945 Great Synagogue looted
Prayerbooks burnt in the street
2 December 1947 Anti-Jewish riots. Many Jews
killed, 150 Jewish homes, 50 shops,
18 synagogues and 5 schools burned
April 1948 Further anti-Jewish riots. Many
Jews in hiding, in fear of their lives

June 1967 Fifty seven Jews killed by
the mob during anti-Jewish riots

November 1950
Thirty Syrian Jews
murdered at sea
by Arab seamen
paid to take them
by boat to Israel
20 bodies washed
ashore at Haifa

T U R K E Y

Kamishliye

Aleppo

River

Euphrates

S Y R I A

Mediterranean
Sea

LEBANON

Damascus

I R A Q

Golan
Heights

Haifa

ISRAEL

R. Jordan

JORDAN

RESTRICTIONS IN FORCE SINCE 1967

1 Jews'right to emigrate is completely forbidden. This
applies even to Jews in Syria who hold foreign passports
2 Jews are forbidden to move more than 3 kilometres
from their place of residence. Those wishing to
travel further must apply for a special permit
3 Identity cards issued to Jews are stamped in red
with the word 'Mussawi' (Jew)
4 Jews are normally subject to a 10 p.m. curfew
5 Jews allowed six years elementary
schooling only
6 Jewish houses in Kamishliye are marked in red
7 Jews barred from jobs in the public service,
in public institutions or in banks
8 Government and military personnel are forbidden
to purchase from Jewish shops
9 Foreigners may not visit the Jewish quarter
unescorted
10 Jews forbidden to own radios or telephones,
or to maintain postal contact with outside world
11 No telephones are installed in Jewish homes
12 The possessions of deceased Jews are confiscated
by the Government. Their heirs must then pay for the
use of the property. If they cannot, it is handed
over to the Palestinian Arabs
13 Only two Jewish schools open in Damascus. Their
directors and most of their teachers are Muslims.
Exams usually ordered to be held on the sabbath

1936-9 Headquarters of anti-Jewish
propaganda, intensified after visit of
Nazi officers from Germany
1938 Jews frequently stabbed on streets
June 1945 A Jewish educationalist murdered
5 August 1949 Bomb thrown in synagogue
on sabbath eve. 12 killed, 26 injured
December 1949 Jewish Community Council
dissolved
8 February 1967 Ministry of Defence
Circular lists 47 Jewish merchants and
forbids army personnel to trade with them
March 1974 Four young Jewesses murdered
while attempting to leave Syria.
Since 1971 at least 50 Jews (men, women
and children) arrested. Many tortured.
Beatings in streets commonplace

© Martin Gilbert 1975

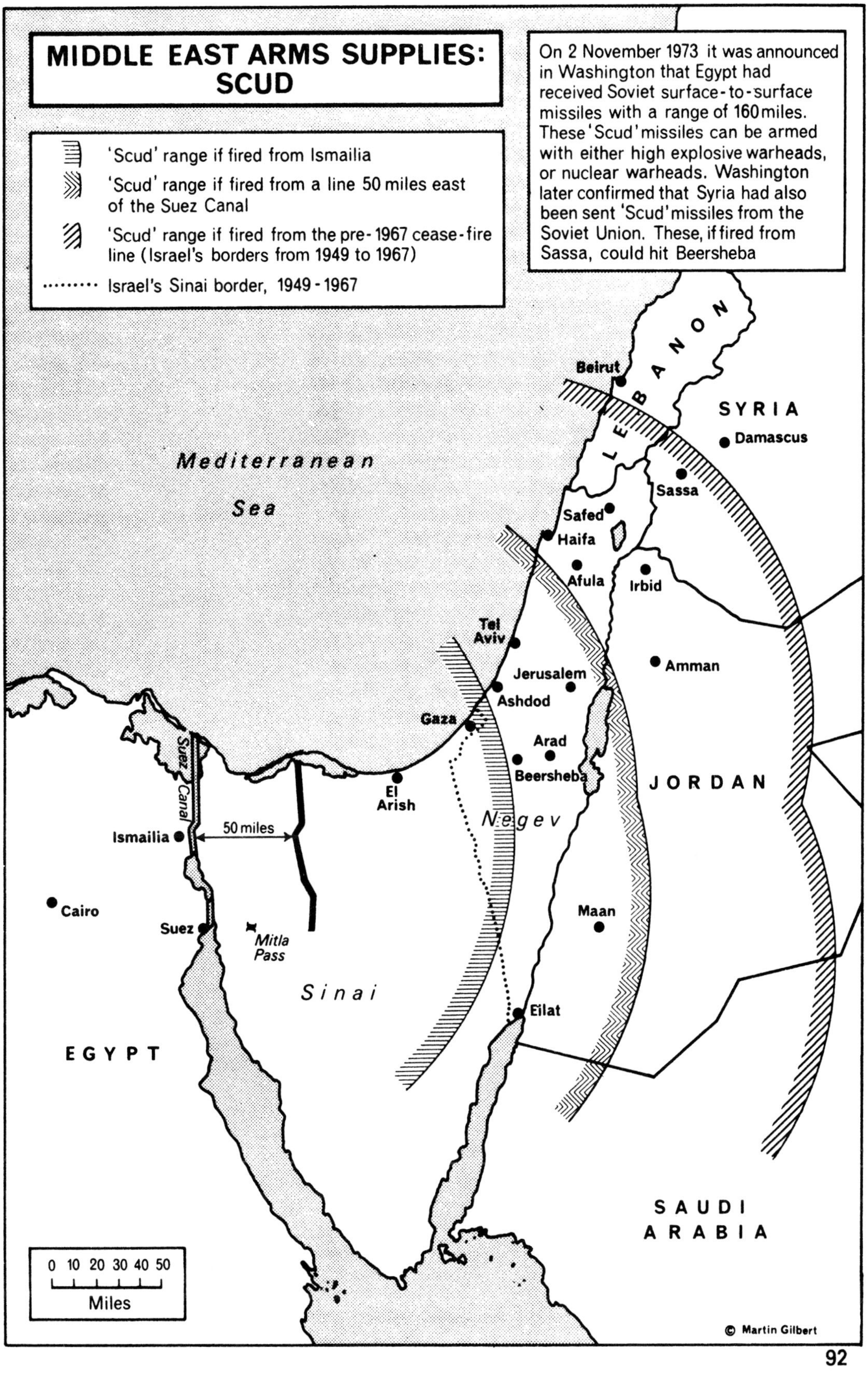

MIDDLE EAST ARMS SUPPLIES: SCUD

'Scud' range if fired from Ismailia

'Scud' range if fired from a line 50 miles east of the Suez Canal

'Scud' range if fired from the pre-1967 cease-fire line (Israel's borders from 1949 to 1967)

········· Israel's Sinai border, 1949-1967

On 2 November 1973 it was announced in Washington that Egypt had received Soviet surface-to-surface missiles with a range of 160 miles. These 'Scud' missiles can be armed with either high explosive warheads, or nuclear warheads. Washington later confirmed that Syria had also been sent 'Scud' missiles from the Soviet Union. These, if fired from Sassa, could hit Beersheba

Mediterranean Sea

L E B A N O N

Beirut

S Y R I A

Damascus

Sassa

Safed

Haifa

Afula

Irbid

Tel Aviv

Jerusalem

Ashdod

Amman

Gaza

Arad

Beersheba

J O R D A N

El Arish

Negev

Suez Canal

50 miles

Ismailia

Maan

Cairo

Suez

Mitla Pass

S i n a i

Eilat

E G Y P T

S A U D I
A R A B I A

0 10 20 30 40 50

Miles

© Martin Gilbert

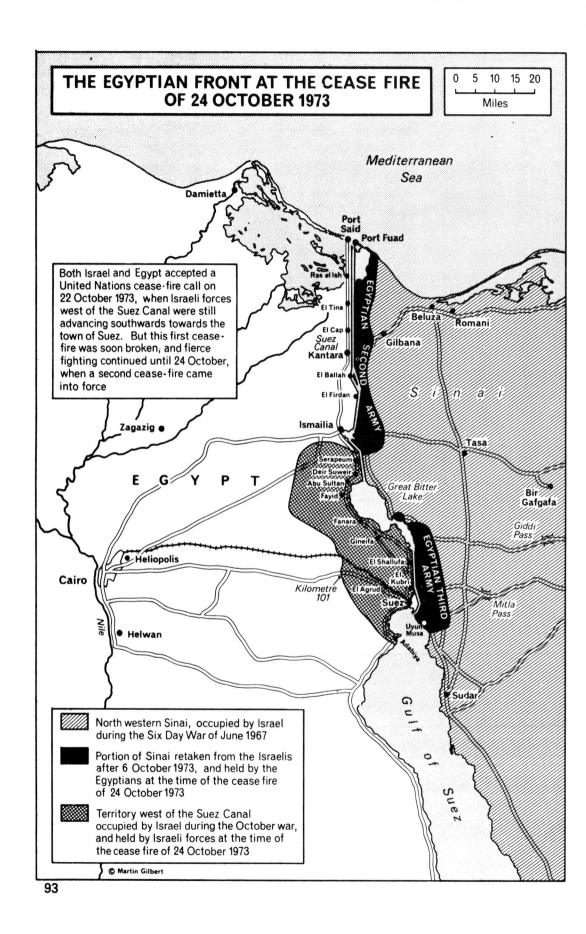

THE EGYPTIAN FRONT AT THE CEASE FIRE OF 24 OCTOBER 1973

0 5 10 15 20
Miles

Both Israel and Egypt accepted a United Nations cease-fire call on 22 October 1973, when Israeli forces west of the Suez Canal were still advancing southwards towards the town of Suez. But this first cease-fire was soon broken, and fierce fighting continued until 24 October, when a second cease-fire came into force

Mediterranean Sea

Damietta

Port Said
Port Fuad

Ras el Ish

El Tina

Beluza   Romani

El Cap
Suez Canal
Kantara

Gilbana

El Ballah

El Firdan

Ismailia

EGYPTIAN SECOND ARMY

Zagazig

S i n a i

Tasa

E G Y P T

Serapeum
Deir Suweir
Abu Sultan
Fayid

Great Bitter Lake

Bir Gafgafa

Giddi Pass

Fanara

Gineifa

EGYPTIAN THIRD ARMY

El Shallufa
El Kubri

Heliopolis

Kilometre 101

El Agrud
Suez

Mitla Pass

Cairo

Nile

Helwan

Uyun Musa

Adabiya

Sudar

Gulf of Suez

North western Sinai, occupied by Israel during the Six Day War of June 1967

Portion of Sinai retaken from the Israelis after 6 October 1973, and held by the Egyptians at the time of the cease fire of 24 October 1973

Territory west of the Suez Canal occupied by Israel during the October war, and held by Israeli forces at the time of the cease fire of 24 October 1973

© Martin Gilbert

93

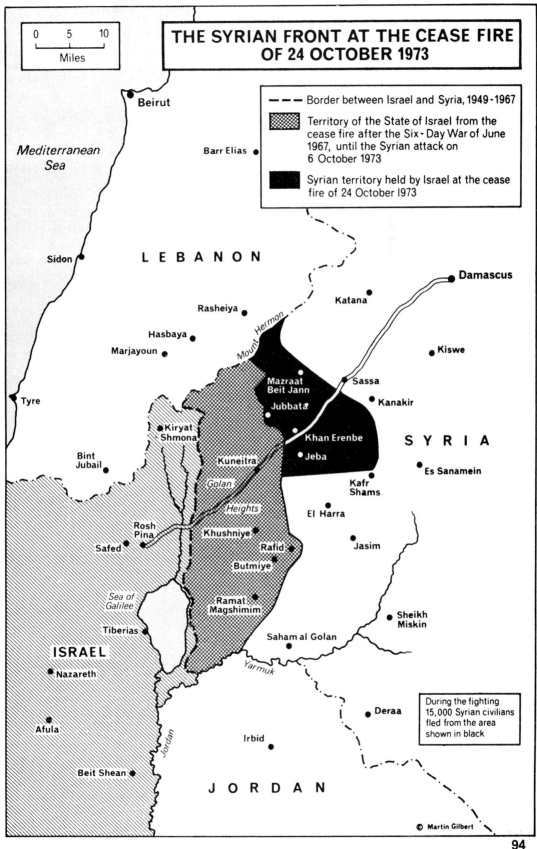

THE SYRIAN FRONT AT THE CEASE FIRE
OF 24 OCTOBER 1973

0   5   10
Miles

– – – Border between Israel and Syria, 1949 - 1967

Territory of the State of Israel from the
cease fire after the Six - Day War of June
1967, until the Syrian attack on
6 October 1973

Syrian territory held by Israel at the cease
fire of 24 October 1973

*Mediterranean
Sea*

Beirut

Barr Elias

Sidon

L E B A N O N

Damascus

Katana

Kiswe

Rasheiya

Mount Hermon

Hasbaya

Marjayoun

Mazraat
Beit Jann

Sassa

Kanakir

Jubbata

Tyre

Kiryat
Shmona

Khan Erenbe

S Y R I A

Jeba

Bint
Jubail

Kuneitra

Es Sanamein

Golan

Kafr
Shams

Heights

El Harra

Rosh
Pina

Khushniye

Jasim

Safed

Rafid

Butmiye

*Sea of
Galilee*

Ramat
Magshimim

Sheikh
Miskin

Tiberias

Saham al Golan

ISRAEL

Yarmuk

Nazareth

Deraa

During the fighting
15,000 Syrian civilians
fled from the area
shown in black

Afula

Jordan

Irbid

Beit Shean

J O R D A N

© Martin Gilbert

94

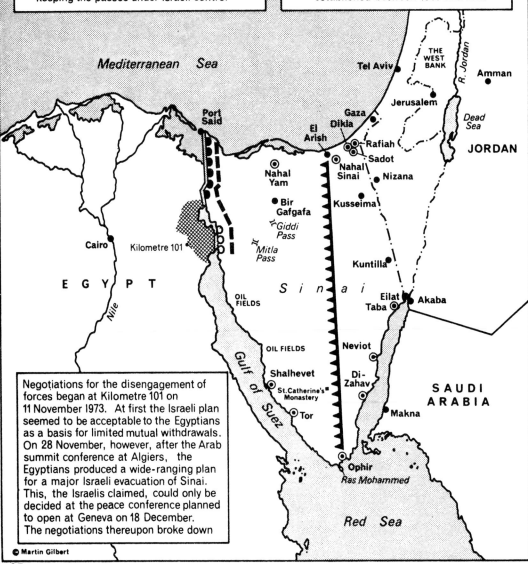

ISRAELI AND EGYPTIAN DISENGAGEMENT PROPOSALS NOVEMBER 1973

0 10 20 30 40 50
Miles

ISRAELI PROPOSAL

1. Israeli forces to withdraw altogether from the west bank of the Suez Canal

2. The Egyptian second army to remain on the east bank of the canal

3. The Egyptian third army to withdraw to the west bank of the canal, but to be replaced by a small Egyptian police force, larger than a token force

4. All Israeli troops to withdraw to a line some ten kilometres west of the 1967-1973 cease-fire line along the canal, keeping the passes under Israeli control

EGYPTIAN PROPOSAL

1. All Israeli forces to be withdrawn east of a line from El Arish to Sharm el-Sheikh, with further withdrawals to be a matter for the Peace Conference

—·— The 1949 cease-fire line, Israel's de-facto border, 1949-1967

⊙ Israeli settlements in Sinai, established between 1967 and 1973

Mediterranean Sea

THE WEST BANK

R. Jordan

Tel Aviv

Amman

Jerusalem

Dead Sea

Port Said

Gaza
Dikla

El Arish

Rafiah
Sadot

JORDAN

Nahal Yam

Nahal Sinai

Nizana

Bir Gafgafa

Kusseima

Cairo

Kilometre 101

Giddi Pass

Mitla Pass

EGYPT

Nile

Sinai

Kuntilla

OIL FIELDS

Eilat
Taba

Akaba

Gulf of Suez

OIL FIELDS

Neviot

SAUDI ARABIA

Shalhevet

St. Catherine's Monastery

Di-Zahav

Tor

Makna

Ophir

Ras Mohammed

Red Sea

Negotiations for the disengagement of forces began at Kilometre 101 on 11 November 1973. At first the Israeli plan seemed to be acceptable to the Egyptians as a basis for limited mutual withdrawals. On 28 November, however, after the Arab summit conference at Algiers, the Egyptians produced a wide-ranging plan for a major Israeli evacuation of Sinai. This, the Israelis claimed, could only be decided at the peace conference planned to open at Geneva on 18 December. The negotiations thereupon broke down

© Martin Gilbert

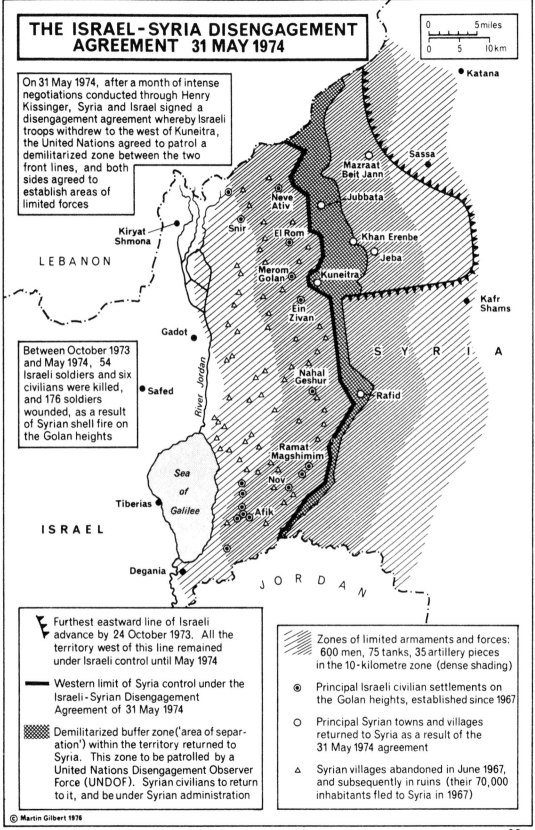

THE ISRAEL-SYRIA DISENGAGEMENT AGREEMENT 31 MAY 1974

On 31 May 1974, after a month of intense negotiations conducted through Henry Kissinger, Syria and Israel signed a disengagement agreement whereby Israeli troops withdrew to the west of Kuneitra, the United Nations agreed to patrol a demilitarized zone between the two front lines, and both sides agreed to establish areas of limited forces

Between October 1973 and May 1974, 54 Israeli soldiers and six civilians were killed, and 176 soldiers wounded, as a result of Syrian shell fire on the Golan heights

Katana

Sassa

Mazraat Beit Jann

Jubbata

Neve Ativ

Snir

El Rom

Khan Erenbe

Jeba

Kiryat Shmona

LEBANON

Merom Golan

Kuneitra

Kafr Shams

Ein Zivan

Gadot

S Y R I A

Nahal Geshur

Safed

Rafid

River Jordan

Ramat Magshimim

Nov

Sea of Galilee

Tiberias

Afik

ISRAEL

Degania

J O R D A N

Furthest eastward line of Israeli advance by 24 October 1973. All the territory west of this line remained under Israeli control until May 1974

Western limit of Syria control under the Israeli-Syrian Disengagement Agreement of 31 May 1974

Demilitarized buffer zone('area of separation') within the territory returned to Syria. This zone to be patrolled by a United Nations Disengagement Observer Force (UNDOF). Syrian civilians to return to it, and be under Syrian administration

Zones of limited armaments and forces: 600 men, 75 tanks, 35 artillery pieces in the 10-kilometre zone (dense shading)

⊙ Principal Israeli civilian settlements on the Golan heights, established since 1967

○ Principal Syrian towns and villages returned to Syria as a result of the 31 May 1974 agreement

△ Syrian villages abandoned in June 1967, and subsequently in ruins (their 70,000 inhabitants fled to Syria in 1967)

© Martin Gilbert 1976

THE ARAB-ISRAELI CONFLICT: AIMS AND OPINIONS NOVEMBER 1973 - MARCH 1974

The war is not over yet. We must admit that our territory has not yet been liberated and we have another fight before us for which we must prepare. **GENERAL GAMASSY, EGYPTIAN CHIEF OF STAFF, 4 MARCH 1974**

On 2 February 1974 the Egyptians announced that they had begun work to open the Suez Canal. But simultaneously with the disengagement of the Egyptian and Israeli forces in Sinai, the Syrians began to bombard Israeli military positions and civilian settlements on the Golan heights. On 3 February 1974 the Syrian Foreign Minister, A.H. Khaddam, announced that Syria was carrying out 'continued and real war of attrition... keeping Israel's reserves on active duty and paralysing its economy'. Throughout March 1974 the Syrians insisted that there could be no negotiations with Israel until Israel withdrew completely from the Golan Heights. On 31 March it was stated in Washington that a 'foreign legion' serving inside Syria included units from Kuwait, Morocco and Saudi Arabia, as well as North Korean pilots and a Cuban armoured brigade with over 100 tanks

The Soviet Union and Syria re-affirm that the establishment of a durable and just peace cannot be achieved in the Middle East unless Israel withdraws from all occupied territories, and the legitimate rights of the Palestinians are restored. **SOVIET FOREIGN MINISTER GROMYKO AND SYRIAN PRESIDENT ASSAD, JOINT STATEMENT 7 MARCH 1974**

Beirut

LEBANON

● Damascus

Kuneitra

SYRIA

The Golan Heights

Mediterranean Sea

Netanya
Tel Aviv

ISRAEL

R. Jordan

JORDAN

Jerusalem

● Amman

Port Said

Gaza

Dead Sea

SAUDI ARABIA

Suez Canal

El Arish

Beersheba

Negev

● Cairo

Bir Gafgafa

Sinai

Suez

Eilat

● Akaba

E G Y P T

Abu Rudeis

Gulf of Suez

Gulf of Eilat

Sharm el-Sheikh

- – - The 1949 cease-fire lines (Israel's borders 1949 - 1967)

▨ Occupied by Israeli forces in June 1967

▼ The front lines at the ceasefire
◄ of October 1973

━ The zone of disengagement in Sinai, March 1974

Palestine is not only part of the Arab homeland but also a basic part of South Syria. We consider it our right and duty to insist that Palestine should remain a free part of our Arab homeland and of our Arab Syrian country. **PRESIDENT ASSAD OF SYRIA, 8 MARCH 1974**

I am not for staying 20 kilometres from the Canal. I do not see in Abu Rudeis, and in oil, the final security line for Israel - although there is oil there - because it means also control of the Suez Canal. I can see all the reasons for wanting control on the Gulf of Eilat, but I cannot think we will have peace with Egypt while we control not only the entrance to Eilat but also the entrance to Suez. **M. DAYAN, ISRAELI DEFENCE MINISTER, 10 MARCH 1974**

We stick to our stand that Israel should withdraw from all Arab territories she occupied since June 1967, and say that there can be no peace in this area without complete withdrawal. We also need not say that Arab Jerusalem, that precious jewel on the forehead of this homeland, will never and under no circumstances come under any sovereignty other than absolute Arab sovereignty. **KING HUSSEIN OF JORDAN, 2 DEC 1973**

For the Syrians, the occupied territories means not only the Golan Heights, but Jerusalem, and even Tel Aviv. **U.S. SECRETARY OF STATE, HENRY KISSINGER, 11 MARCH 1974**

....we will not descend from the Golan, we will not partition Jerusalem, we will not return Sharm el-Sheikh, and we will not agree that the distance between Netanya and the border shall be 18 kilometres.... But if we want a Jewish State we have to be prepared to compromise on territory. **GOLDA MEIR, ISRAELI PRIME MINISTER, 29 DEC 1973**

.... the talk in Israel about a demilitarized Sinai should stop. If they want a demilitarized Sinai, I shall be asking for a demilitarized Israel. **PRESIDENT SADAT OF EGYPT, 'TIME' MAGAZINE, 25 MARCH 1974**

© Martin Gilbert

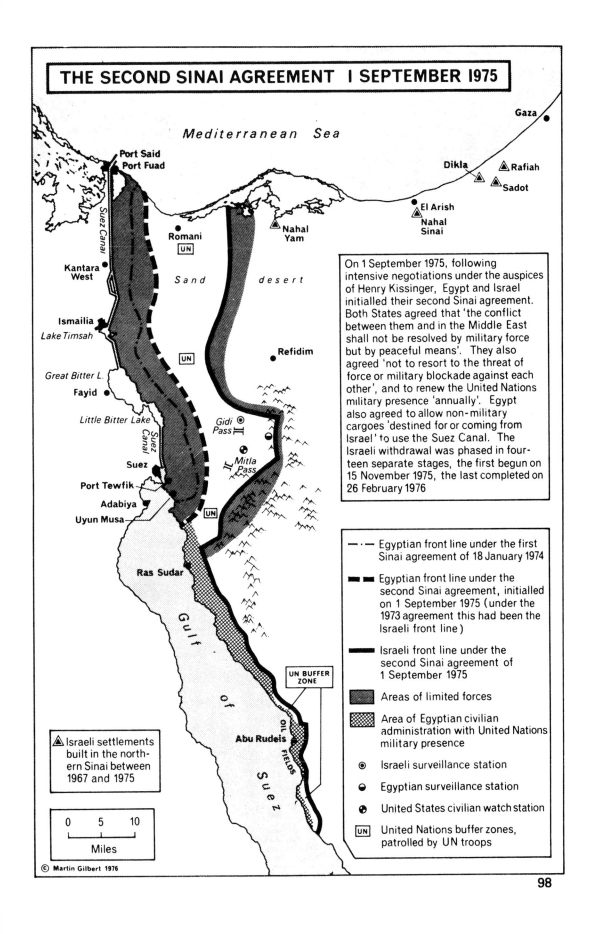

THE SECOND SINAI AGREEMENT 1 SEPTEMBER 1975

Mediterranean Sea

Gaza

Port Said
Port Fuad

Dikla ▲Rafiah
▲Sadot

▲El Arish
Nahal Sinai

Nahal Yam

Romani
UN

Kantara West

Sand   desert

Ismailia
Lake Timsah

Refidim

Great Bitter L.

UN

Fayid

Little Bitter Lake

Gidi Pass ◉

Mitla Pass

Suez

Port Tewfik

Adabiya

Uyun Musa

UN

Ras Sudar

Gulf   of

UN BUFFER ZONE

Abu Rudeis

OIL FIELDS

Suez

On 1 September 1975, following intensive negotiations under the auspices of Henry Kissinger, Egypt and Israel initialled their second Sinai agreement. Both States agreed that 'the conflict between them and in the Middle East shall not be resolved by military force but by peaceful means'. They also agreed 'not to resort to the threat of force or military blockade against each other', and to renew the United Nations military presence 'annually'. Egypt also agreed to allow non-military cargoes 'destined for or coming from Israel' to use the Suez Canal. The Israeli withdrawal was phased in fourteen separate stages, the first begun on 15 November 1975, the last completed on 26 February 1976

—·— Egyptian front line under the first Sinai agreement of 18 January 1974

▬ ▬ Egyptian front line under the second Sinai agreement, initialled on 1 September 1975 (under the 1973 agreement this had been the Israeli front line)

▬▬ Israeli front line under the second Sinai agreement of 1 September 1975

▒ Areas of limited forces

▤ Area of Egyptian civilian administration with United Nations military presence

◉ Israeli surveillance station

◒ Egyptian surveillance station

◈ United States civilian watch station

UN United Nations buffer zones, patrolled by UN troops

▲ Israeli settlements built in the northern Sinai between 1967 and 1975

0   5   10
Miles

© Martin Gilbert 1976

98

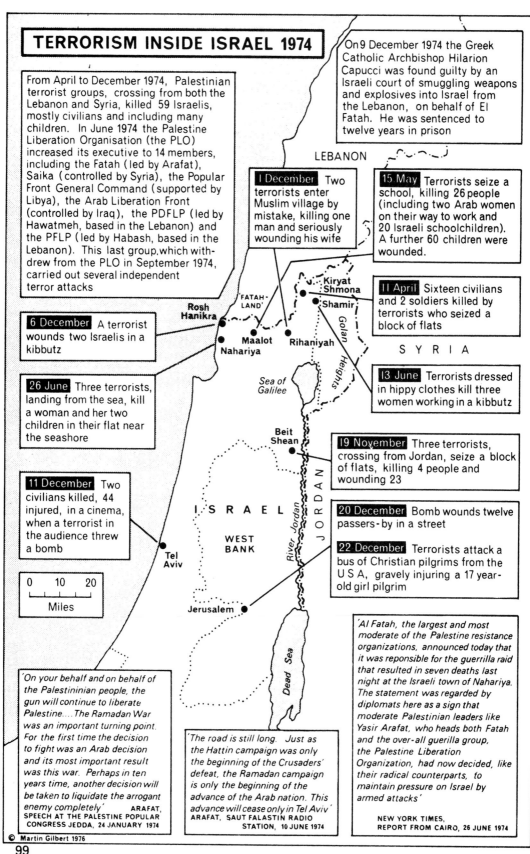

TERRORISM INSIDE ISRAEL 1974

On 9 December 1974 the Greek Catholic Archbishop Hilarion Capucci was found guilty by an Israeli court of smuggling weapons and explosives into Israel from the Lebanon, on behalf of El Fatah. He was sentenced to twelve years in prison

From April to December 1974, Palestinian terrorist groups, crossing from both the Lebanon and Syria, killed 59 Israelis, mostly civilians and including many children. In June 1974 the Palestine Liberation Organisation (the PLO) increased its executive to 14 members, including the Fatah (led by Arafat), Saika (controlled by Syria), the Popular Front General Command (supported by Libya), the Arab Liberation Front (controlled by Iraq), the PDFLP (led by Hawatmeh, based in the Lebanon) and the PFLP (led by Habash, based in the Lebanon). This last group, which withdrew from the PLO in September 1974, carried out several independent terror attacks

LEBANON

I December Two terrorists enter Muslim village by mistake, killing one man and seriously wounding his wife

15 May Terrorists seize a school, killing 26 people (including two Arab women on their way to work and 20 Israeli schoolchildren). A further 60 children were wounded.

11 April Sixteen civilians and 2 soldiers killed by terrorists who seized a block of flats

Kiryat Shmona
Shamir

'FATAH-LAND'

Golan Heights

Rosh Hanikra

6 December A terrorist wounds two Israelis in a kibbutz

Maalot
Nahariya
Rihaniyah

S Y R I A

13 June Terrorists dressed in hippy clothes kill three women working in a kibbutz

Sea of Galilee

26 June Three terrorists, landing from the sea, kill a woman and her two children in their flat near the seashore

Beit Shean

19 November Three terrorists, crossing from Jordan, seize a block of flats, killing 4 people and wounding 23

11 December Two civilians killed, 44 injured, in a cinema, when a terrorist in the audience threw a bomb

J O R D A N

River Jordan

20 December Bomb wounds twelve passers-by in a street

I S R A E L

WEST BANK

22 December Terrorists attack a bus of Christian pilgrims from the U S A, gravely injuring a 17 year-old girl pilgrim

Tel Aviv

0 10 20
Miles

Jerusalem

Dead Sea

'Al Fatah, the largest and most moderate of the Palestine resistance organizations, announced today that it was reponsible for the guerrilla raid that resulted in seven deaths last night at the Israeli town of Nahariya. The statement was regarded by diplomats here as a sign that moderate Palestinian leaders like Yasir Arafat, who heads both Fatah and the over-all guerilla group, the Palestine Liberation Organization, had now decided, like their radical counterparts, to maintain pressure on Israel by armed attacks'

'On your behalf and on behalf of the Palestininian people, the gun will continue to liberate Palestine....The Ramadan War was an important turning point. For the first time the decision to fight was an Arab decision and its most important result was this war. Perhaps in ten years time, another decision will be taken to liquidate the arrogant enemy completely' **ARAFAT, SPEECH AT THE PALESTINE POPULAR CONGRESS JEDDA, 24 JANUARY 1974**

'The road is still long. Just as the Hattin campaign was only the beginning of the Crusaders' defeat, the Ramadan campaign is only the beginning of the advance of the Arab nation. This advance will cease only in Tel Aviv' **ARAFAT, SAUT FALASTIN RADIO STATION, 10 JUNE 1974**

NEW YORK TIMES, REPORT FROM CAIRO, 26 JUNE 1974

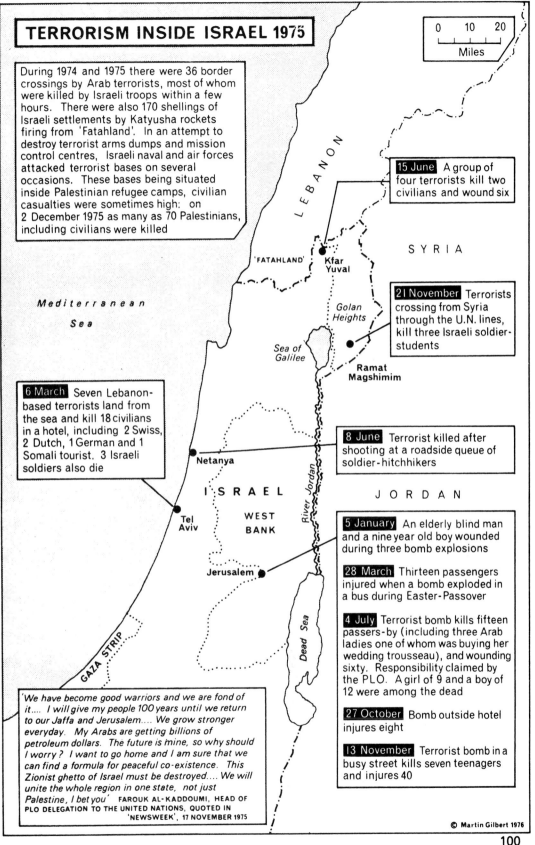

TERRORISM INSIDE ISRAEL 1975

During 1974 and 1975 there were 36 border crossings by Arab terrorists, most of whom were killed by Israeli troops within a few hours. There were also 170 shellings of Israeli settlements by Katyusha rockets firing from 'Fatahland'. In an attempt to destroy terrorist arms dumps and mission control centres, Israeli naval and air forces attacked terrorist bases on several occasions. These bases being situated inside Palestinian refugee camps, civilian casualties were sometimes high: on 2 December 1975 as many as 70 Palestinians, including civilians were killed

15 June A group of four terrorists kill two civilians and wound six

21 November Terrorists crossing from Syria through the U.N. lines, kill three Israeli soldier-students

6 March Seven Lebanon-based terrorists land from the sea and kill 18 civilians in a hotel, including 2 Swiss, 2 Dutch, 1 German and 1 Somali tourist. 3 Israeli soldiers also die

8 June Terrorist killed after shooting at a roadside queue of soldier-hitchhikers

5 January An elderly blind man and a nine year old boy wounded during three bomb explosions

28 March Thirteen passengers injured when a bomb exploded in a bus during Easter-Passover

4 July Terrorist bomb kills fifteen passers-by (including three Arab ladies one of whom was buying her wedding trousseau), and wounding sixty. Responsibility claimed by the PLO. A girl of 9 and a boy of 12 were among the dead

27 October Bomb outside hotel injures eight

13 November Terrorist bomb in a busy street kills seven teenagers and injures 40

LEBANON

SYRIA

'FATAHLAND' Kfar Yuval

Golan Heights

Sea of Galilee

Ramat Magshimim

Mediterranean Sea

Netanya

I S R A E L

WEST BANK

River Jordan

J O R D A N

Tel Aviv

Jerusalem

Dead Sea

GAZA STRIP

'We have become good warriors and we are fond of it.... I will give my people 100 years until we return to our Jaffa and Jerusalem.... We grow stronger everyday. My Arabs are getting billions of petroleum dollars. The future is mine, so why should I worry? I want to go home and I am sure that we can find a formula for peaceful co-existence. This Zionist ghetto of Israel must be destroyed.... We will unite the whole region in one state, not just Palestine, I bet you' FAROUK AL-KADDOUMI, HEAD OF PLO DELEGATION TO THE UNITED NATIONS, QUOTED IN 'NEWSWEEK', 17 NOVEMBER 1975

© Martin Gilbert 1976

0   10   20
Miles

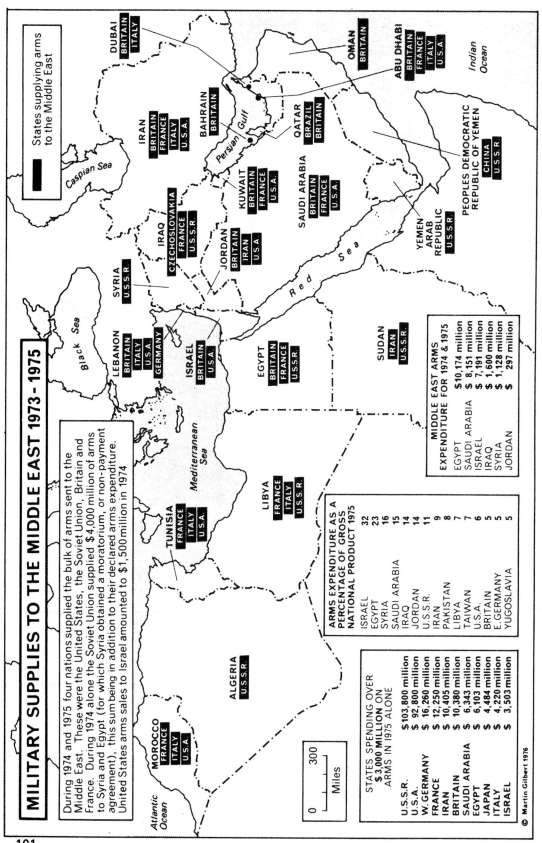

MILITARY SUPPLIES TO THE MIDDLE EAST 1973 - 1975

During 1974 and 1975 four nations supplied the bulk of arms sent to the Middle East. These were the United States, the Soviet Union, Britain and France. During 1974 alone the Soviet Union supplied $4,000 million of arms to Syria and Egypt (for which Syria obtained a moratorium, or non-payment agreement), this sum being in addition to their declared arms expenditure. United States arms sales to Israel amounted to $1,500 million in 1974

States supplying arms to the Middle East

DUBAI
BRITAIN
ITALY

OMAN
BRITAIN

ABU DHABI
BRITAIN
FRANCE
ITALY
U.S.A.

IRAN
BRITAIN
FRANCE
ITALY
U.S.A.

BAHRAIN
BRITAIN

QATAR
BRAZIL
BRITAIN

SAUDI ARABIA
BRITAIN
FRANCE
U.S.A.

KUWAIT
BRITAIN
FRANCE
U.S.A.

PEOPLES DEMOCRATIC REPUBLIC OF YEMEN
CHINA
U.S.S.R.

IRAQ
CZECHOSLOVAKIA
FRANCE
U.S.S.R.

JORDAN
BRITAIN
IRAN
U.S.

YEMEN ARAB REPUBLIC
U.S.S.R.

SYRIA
U.S.S.R.

LEBANON
ITALY
U.S.A.

ISRAEL
BRITAIN
GERMANY
U.S.A.

EGYPT
BRITAIN
FRANCE
U.S.S.R.

SUDAN
IRAN
U.S.S.R.

LIBYA
FRANCE
ITALY
U.S.S.R.

TUNISIA
FRANCE
ITALY
U.S.A.

ALGERIA
U.S.S.R.

MOROCCO
FRANCE
ITALY
U.S.A.

Caspian Sea

Black Sea

Mediterranean Sea

Red Sea

Persian Gulf

Indian Ocean

Atlantic Ocean

MIDDLE EAST ARMS EXPENDITURE FOR 1974 & 1975	
EGYPT	$10,174 million
SAUDI ARABIA	$ 8,151 million
ISRAEL	$ 7,191 million
IRAQ	$ 1,600 million
SYRIA	$ 1,128 million
JORDAN	$ 297 million

ARMS EXPENDITURE AS A PERCENTAGE OF GROSS NATIONAL PRODUCT 1975	
ISRAEL	32
EGYPT	23
SYRIA	16
SAUDI ARABIA	15
IRAQ	14
JORDAN	14
U.S.S.R.	11
IRAN	9
PAKISTAN	8
LIBYA	7
TAIWAN	7
U.S.A.	6
BRITAIN	5
E. GERMANY	5
YUGOSLAVIA	5

STATES SPENDING OVER $3,000 MILLION ON ARMS IN 1975 ALONE	
U.S.S.R.	$103,800 million
U.S.A.	$ 92,800 million
W. GERMANY	$ 16,260 million
FRANCE	$ 12,250 million
IRAN	$ 10,405 million
BRITAIN	$ 10,380 million
SAUDI ARABIA	$ 6,343 million
EGYPT	$ 6,103 million
JAPAN	$ 4,484 million
ITALY	$ 4,220 million
ISRAEL	$ 3,503 million

0	300

Miles

© **Martin Gilbert 1976**

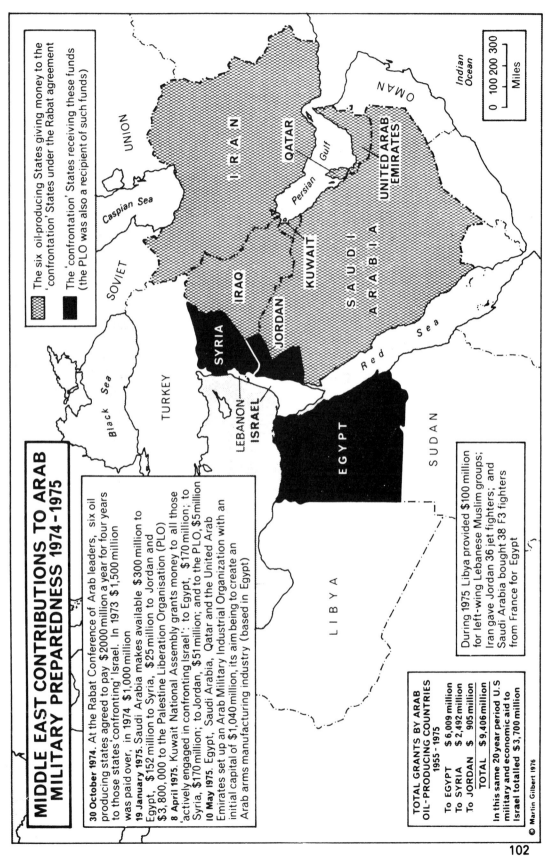

MIDDLE EAST CONTRIBUTIONS TO ARAB MILITARY PREPAREDNESS 1974–1975

The six oil-producing States giving money to the 'confrontation' States under the Rabat agreement

The 'confrontation' States receiving these funds (the PLO was also a recipient of such funds)

30 October 1974. At the Rabat Conference of Arab leaders, six oil producing states agreed to pay $2000 million a year for four years to those states confronting Israel. In 1973 $1,500 million was paid over, in 1974 $1,000 million

19 January 1975. Saudi Arabia makes available $300 million to Egypt, $152 million to Syria, $25 million to Jordan and $3,800,000 to the Palestine Liberation Organisation (PLO)

8 April 1975. Kuwait National Assembly grants money to all those 'actively engaged in confronting Israel': to Egypt, $170 million; to Syria, $170 million; to Jordan, $51 million; and to the PLO, $5 million

10 May 1975. Egypt, Saudi Arabia, Qatar and the United Arab Emirates set up an Arab Military Industrial Organization with an initial capital of $1,040 million, its aim being to create an Arab arms manufacturing industry (based in Egypt)

TOTAL GRANTS BY ARAB
OIL-PRODUCING COUNTRIES
1955 – 1975

To EGYPT	$ 6,009 million
To SYRIA	$ 2,492 million
To JORDAN	$ 905 million
TOTAL	$ 9,406 million

In this same 20 year period U.S military and economic aid to Israel totalled $3,700 million

During 1975 Libya provided $100 million for left-wing Lebanese Muslim groups; Iran gave Jordan 36 jet fighters; and Saudi Arabia bought 38 F3 fighters from France for Egypt

© Martin Gilbert 1976

102

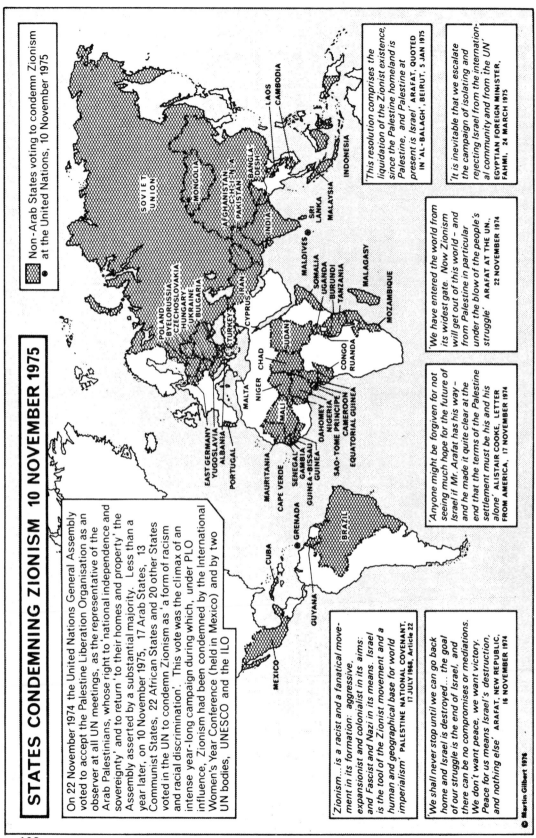

STATES CONDEMNING ZIONISM 10 NOVEMBER 1975

On 22 November 1974 the United Nations General Assembly voted to accept the Palestine Liberation Organisation as an observer at all UN meetings, as the representative of the Arab Palestinians, whose right to 'national independence and sovereignty' and to return 'to their homes and property' the Assembly asserted by a substantial majority. Less than a year later, on 10 November 1975, 17 Arab States, 13 Communist States, 22 African States and 20 other States voted in the UN to condemn Zionism as 'a form of racism and racial discrimination'. This vote was the climax of an intense year-long campaign during which, under PLO influence, Zionism had been condemned by the International Women's Year Conference (held in Mexico) and by two UN bodies, UNESCO and the ILO

'Zionism....is a racist and a fanatical movement in its formation; aggressive,
expansionist and colonialist in its aims;
and Fascist and Nazi in its means. Israel
is the tool of the Zionist movement and a
human and geographical base for world
imperialism' PALESTINE NATIONAL COVENANT,
17 JULY 1968, Article 22

'We shall never stop until we can go back
home and Israel is destroyed.... the goal
of our struggle is the end of Israel, and
there can be no compromises or mediations.
We don't want peace, we want victory.
Peace for us means Israel's destruction,
and nothing else' ARAFAT, NEW REPUBLIC,
16 NOVEMBER 1974

'Anyone might be forgiven for not
seeing much hope for the future of
Israel if Mr. Arafat has his way –
and he made it quite clear at the
end that the terms of the Palestine
settlement must be his and his
alone' ALISTAIR COOKE, LETTER
FROM AMERICA, 17 NOVEMBER 1974

'We have entered the world from
its widest gate. Now Zionism
will get out of this world – and
from Palestine in particular
under the blow of the people's
struggle' ARAFAT AT THE UN,
22 NOVEMBER 1974

'This resolution comprises the
liquidation of the Zionist existence,
since the Palestine homeland is
Palestine, and Palestine at
present is Israel' ARAFAT, QUOTED
IN 'AL-BALAGH', BEIRUT, 5 JAN 1975

'It is inevitable that we escalate
the campaign of isolating and
rejecting Israel from the international community and from the UN'
EGYPTIAN FOREIGN MINISTER,
FAHMI, 24 MARCH 1975

Non-Arab States voting to condemn Zionism
at the United Nations, 10 November 1975

© Martin Gilbert 1976

103

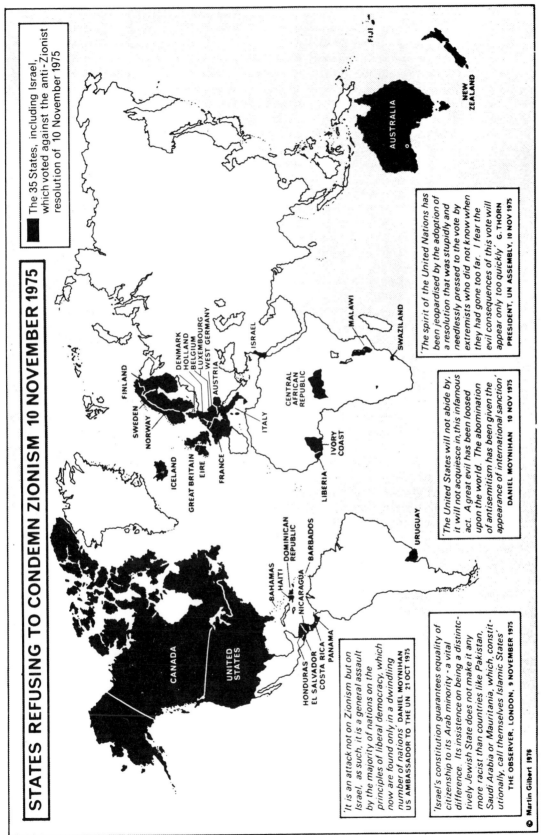

STATES REFUSING TO CONDEMN ZIONISM 10 NOVEMBER 1975

■ The 35 States, including Israel, which voted against the anti-Zionist resolution of 10 November 1975

'It is an attack not on Zionism but on Israel, as such, it is a general assault by the majority of nations on the principles of liberal democracy, which now are found only in a dwindling number of nations' DANIEL MOYNIHAN US AMBASSADOR TO THE UN 21 OCT 1975

'Israel's constitution guarantees equality of citizenship to its Arab minority - a vital difference. Its insistence on being a distinctively Jewish State does not make it any more racist than countries like Pakistan, Saudi Arabia or Mauritania, which constitutionally, call themselves Islamic States' THE OBSERVER, LONDON, 9 NOVEMBER 1975

'The United States will not abide by, it will not acquiesce in, this infamous act. A great evil has been loosed upon the world. The abomination of antisemitism has been given the appearance of international sanction' DANIEL MOYNIHAN 10 NOV 1975

'The spirit of the United Nations has been jeopardised by the adoption of a resolution that was stupidly and needlessly pressed to the vote by extremists who did not know when they had gone too far. I fear the evil consequences of this vote will appear only too quickly' G. THORN PRESIDENT, UN ASSEMBLY, 10 NOV 1975

© Martin Gilbert 1976

CANADA
UNITED STATES
BAHAMAS
HAITI
DOMINICAN REPUBLIC
NICARAGUA
BARBADOS
HONDURAS
EL SALVADOR
COSTA RICA
PANAMA
URUGUAY

ICELAND
GREAT BRITAIN
EIRE
FRANCE
FINLAND
SWEDEN
NORWAY
DENMARK
HOLLAND
BELGIUM
LUXEMBOURG
WEST GERMANY
AUSTRIA
ITALY
ISRAEL
LIBERIA
IVORY COAST
CENTRAL AFRICAN REPUBLIC
MALAWI
SWAZILAND

AUSTRALIA
NEW ZEALAND
FIJI

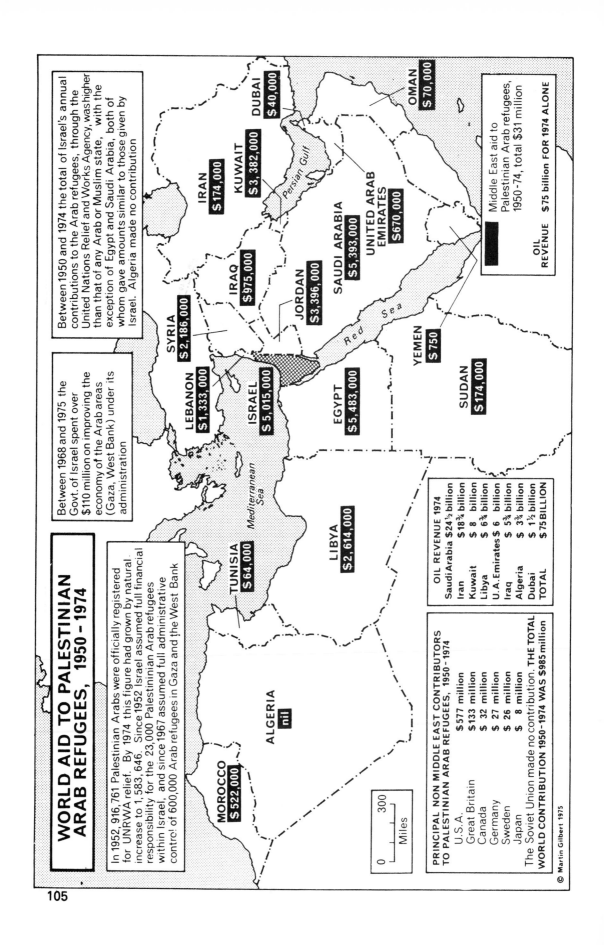

WORLD AID TO PALESTINIAN ARAB REFUGEES, 1950 – 1974

In 1952, 916,761 Palestinian Arabs were officially registered for UNRWA relief. By 1974 this figure had grown by natural increase to 1,583,646. Since 1952 Israel assumed full financial responsibility for the 23,000 Palestinian Arab refugees within Israel, and since 1967 assumed full administrative control of 600,000 Arab refugees in Gaza and the West Bank

Between 1968 and 1975 the Govt. of Israel spent over $110 million on improving the economy of the Arab areas (Gaza, West Bank) under its administration

Between 1950 and 1974 the total of Israel's annual contributions to the Arab refugees, through the United Nations Relief and Works Agency, was higher than that of any Arab or Muslim state, with the exception of Egypt and Saudi Arabia, both of whom gave amounts similar to those given by Israel. Algeria made no contribution

MOROCCO $522,000

ALGERIA nil

TUNISIA $64,000

LIBYA $2,614,000

Mediterranean Sea

LEBANON $1,333,000

ISRAEL $5,015,000

SYRIA $2,186,000

IRAQ $975,000

JORDAN $3,396,000

EGYPT $5,483,000

Red Sea

SUDAN $174,000

YEMEN $750

SAUDI ARABIA $5,393,000

UNITED ARAB EMIRATES $670,000

KUWAIT $3,382,000

Persian Gulf

IRAN $174,000

DUBAI $40,000

OMAN $70,000

Middle East aid to Palestinian Arab refugees, 1950-74, total $31 million

| OIL REVENUE | $75 billion FOR 1974 ALONE |

OIL REVENUE 1974	
Saudi Arabia	$24½ billion
Iran	$18¾ billion
Kuwait	$ 8 billion
Libya	$ 6¾ billion
U.A. Emirates	$ 6 billion
Iraq	$ 5¾ billion
Algeria	$ 3¾ billion
Dubai	$ 1½ billion
TOTAL	$75 BILLION

PRINCIPAL NON MIDDLE EAST CONTRIBUTORS TO PALESTINIAN ARAB REFUGEES, 1950-1974	
U.S.A.	$577 million
Great Britain	$133 million
Canada	$ 32 million
Germany	$ 27 million
Sweden	$ 26 million
Japan	$ 8 million

The Soviet Union made no contribution. THE TOTAL WORLD CONTRIBUTION 1950-1974 WAS $985 million

0	300

Miles

© Martin Gilbert 1975

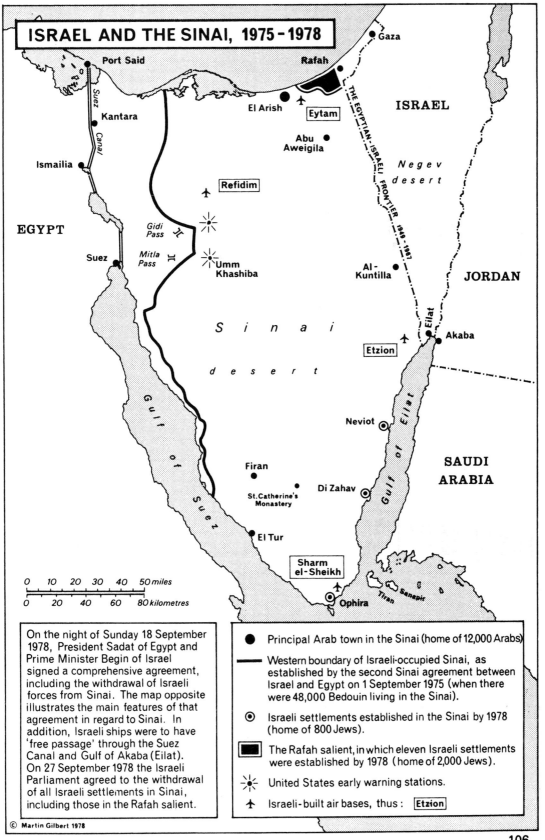

ISRAEL AND THE SINAI, 1975 - 1978

Gaza

Port Said

Rafah

Suez Canal

Kantara

El Arish

Eytam

ISRAEL

Abu Aweigila

THE EGYPTIAN-ISRAELI FRONTIER 1949-1967

Ismailia

Refidim

N e g e v desert

EGYPT

Gidi Pass

Mitla Pass

Umm Khashiba

Al-Kuntilla

JORDAN

Suez

S i n a i

Etzion

Eilat

Akaba

d e s e r t

Gulf of Suez

Gulf of Eilat

Neviot

SAUDI ARABIA

Firan

St. Catherine's Monastery

Di Zahav

El Tur

Sharm el-Sheikh

Sanapir

Tiran

Ophira

0 10 20 30 40 50 miles

0 20 40 60 80 kilometres

On the night of Sunday 18 September 1978, President Sadat of Egypt and Prime Minister Begin of Israel signed a comprehensive agreement, including the withdrawal of Israeli forces from Sinai. The map opposite illustrates the main features of that agreement in regard to Sinai. In addition, Israeli ships were to have 'free passage' through the Suez Canal and Gulf of Akaba (Eilat). On 27 September 1978 the Israeli Parliament agreed to the withdrawal of all Israeli settlements in Sinai, including those in the Rafah salient.

●   Principal Arab town in the Sinai (home of 12,000 Arabs)

—   Western boundary of Israeli-occupied Sinai, as established by the second Sinai agreement between Israel and Egypt on 1 September 1975 (when there were 48,000 Bedouin living in the Sinai).

◉   Israeli settlements established in the Sinai by 1978 (home of 800 Jews).

■   The Rafah salient, in which eleven Israeli settlements were established by 1978 (home of 2,000 Jews).

☀   United States early warning stations.

✈   Israeli-built air bases, thus :   Etzion

© **Martin Gilbert 1978**

106

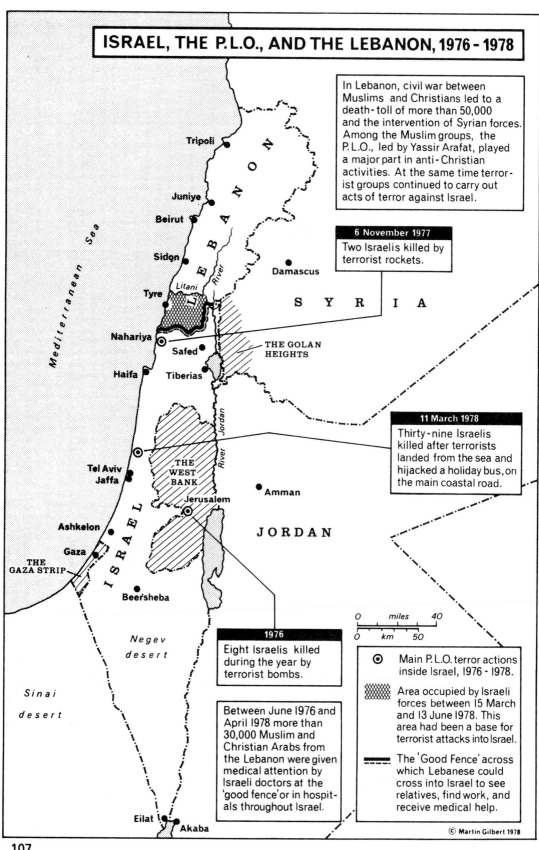

ISRAEL, THE P.L.O., AND THE LEBANON, 1976 - 1978

In Lebanon, civil war between Muslims and Christians led to a death-toll of more than 50,000 and the intervention of Syrian forces. Among the Muslim groups, the P.L.O., led by Yassir Arafat, played a major part in anti-Christian activities. At the same time terrorist groups continued to carry out acts of terror against Israel.

6 November 1977
Two Israelis killed by terrorist rockets.

11 March 1978
Thirty-nine Israelis killed after terrorists landed from the sea and hijacked a holiday bus, on the main coastal road.

1976
Eight Israelis killed during the year by terrorist bombs.

Between June 1976 and April 1978 more than 30,000 Muslim and Christian Arabs from the Lebanon were given medical attention by Israeli doctors at the 'good fence' or in hospitals throughout Israel.

Tripoli

Juniye

Beirut

Sidon

L E B A N O N

Litani

Tyre

Nahariya

Safed

Haifa

Tiberias

Damascus

S Y R I A

River Litani

THE GOLAN HEIGHTS

River Jordan

Mediterranean Sea

Tel Aviv
Jaffa

THE WEST BANK

Jerusalem

Amman

J O R D A N

Ashkelon

Gaza

THE GAZA STRIP

I S R A E L

Beer'sheba

Negev desert

Sinai desert

Eilat
Akaba

| 0 | miles | 40 |
| 0 | km | 50 |

⊙   Main P.L.O. terror actions inside Israel, 1976 - 1978.

▦   Area occupied by Israeli forces between 15 March and 13 June 1978. This area had been a base for terrorist attacks into Israel.

━━   The 'Good Fence' across which Lebanese could cross into Israel to see relatives, find work, and receive medical help.

© Martin Gilbert 1978

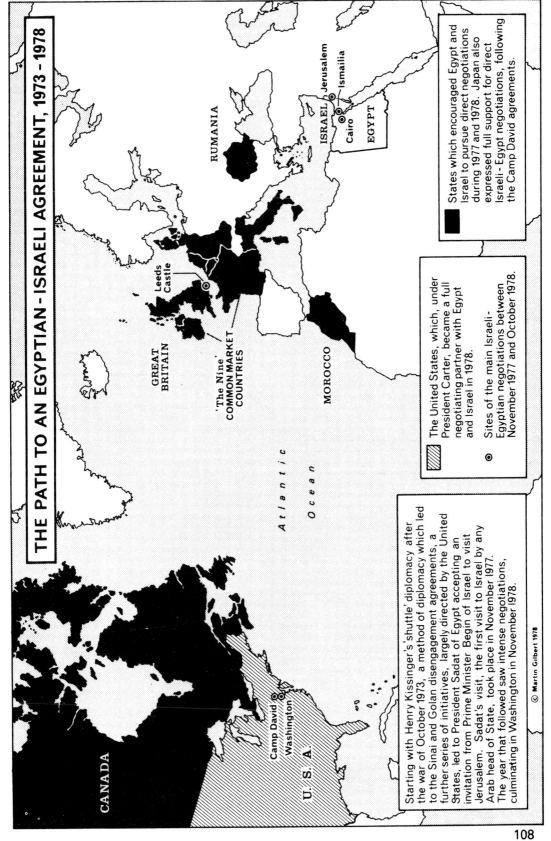

THE PATH TO AN EGYPTIAN - ISRAELI AGREEMENT, 1973 - 1978

RUMANIA

ISRAEL Jerusalem
Cairo Ismailia
EGYPT

Leeds Castle

GREAT BRITAIN

'The Nine' COMMON MARKET COUNTRIES

Atlantic Ocean

MOROCCO

CANADA

Camp David
Washington
U.S.A.

States which encouraged Egypt and Israel to pursue direct negotiations during 1977 and 1978. Japan also expressed full support for direct Israeli-Egypt negotiations, following the Camp David agreements.

The United States, which, under President Carter, became a full negotiating partner with Egypt and Israel in 1978.

⊙ Sites of the main Israeli-Egyptian negotiations between November 1977 and October 1978.

Starting with Henry Kissinger's 'shuttle' diplomacy after the war of October 1973, a method of diplomacy which led to the Sinai and Golan disengagement agreements, a further series of initiatives, largely directed by the United States, led to President Sadat of Egypt accepting an invitation from Prime Minister Begin of Israel to visit Jerusalem. Sadat's visit, the first visit to Israel by any Arab head of State, took place in November 1977. The year that followed saw intense negotiations, culminating in Washington in November 1978.

© Martin Gilbert 1978

108

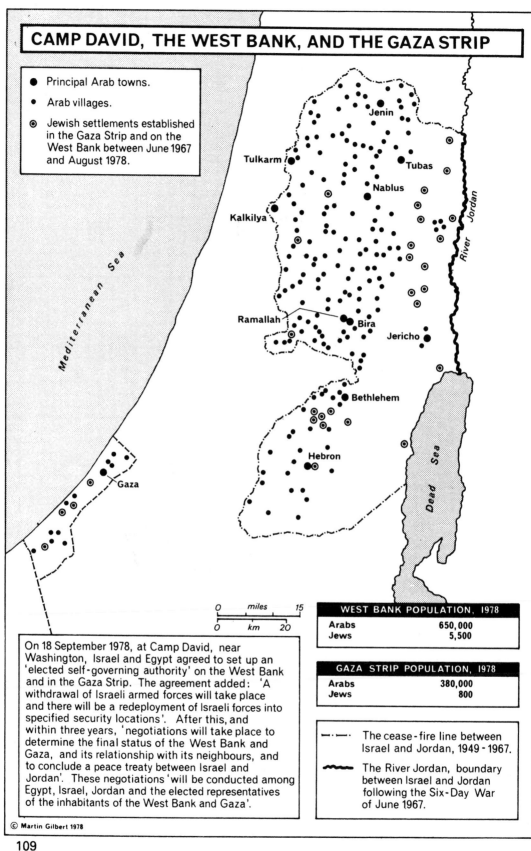

CAMP DAVID, THE WEST BANK, AND THE GAZA STRIP

- ● Principal Arab towns.
- · Arab villages.
- ◉ Jewish settlements established in the Gaza Strip and on the West Bank between June 1967 and August 1978.

Mediterranean Sea

Jenin

Tulkarm

Tubas

Nablus

Kalkilya

River Jordan

Ramallah
Bira

Jericho

Bethlehem

Dead Sea

Hebron

Gaza

	miles	
0		15
0	km	20

On 18 September 1978, at Camp David, near Washington, Israel and Egypt agreed to set up an 'elected self-governing authority' on the West Bank and in the Gaza Strip. The agreement added: 'A withdrawal of Israeli armed forces will take place and there will be a redeployment of Israeli forces into specified security locations'. After this, and within three years, 'negotiations will take place to determine the final status of the West Bank and Gaza, and its relationship with its neighbours, and to conclude a peace treaty between Israel and Jordan'. These negotiations 'will be conducted among Egypt, Israel, Jordan and the elected representatives of the inhabitants of the West Bank and Gaza'.

WEST BANK POPULATION, 1978	
Arabs	650,000
Jews	5,500

GAZA STRIP POPULATION, 1978	
Arabs	380,000
Jews	800

—·—·— The cease-fire line between Israel and Jordan, 1949-1967.

〰〰 The River Jordan, boundary between Israel and Jordan following the Six-Day War of June 1967.

© Martin Gilbert 1978

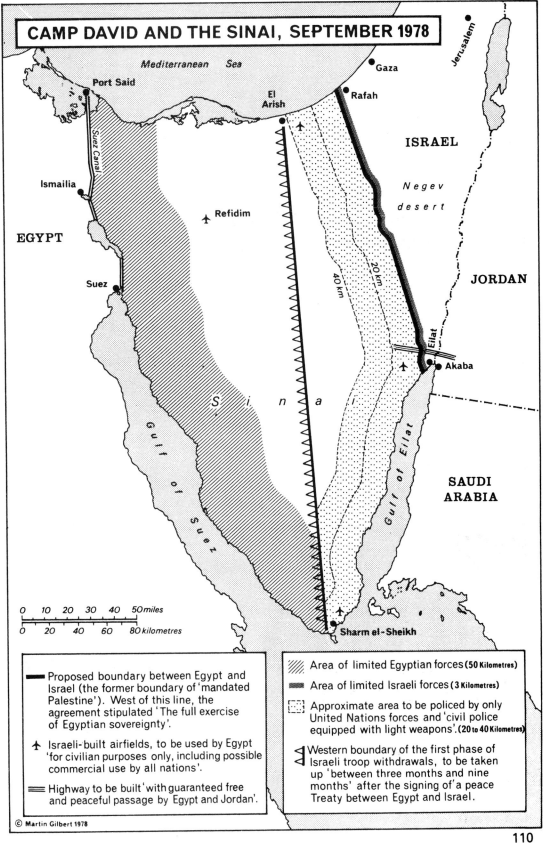

CAMP DAVID AND THE SINAI, SEPTEMBER 1978

Mediterranean Sea

Jerusalem

Gaza

Port Said

El Arish

Rafah

ISRAEL

Ismailia

N e g e v

Suez Canal

EGYPT

Refidim

d e s e r t

20 km

40 km

JORDAN

Suez

S i n a i

Eilat

Akaba

SAUDI ARABIA

Gulf of Suez

Gulf of Eilat

0	10	20	30	40	50 miles
0	20	40	60	80 kilometres	

Sharm el-Sheikh

── Proposed boundary between Egypt and Israel (the former boundary of 'mandated Palestine'). West of this line, the agreement stipulated 'The full exercise of Egyptian sovereignty'.

✈ Israeli-built airfields, to be used by Egypt 'for civilian purposes only, including possible commercial use by all nations'.

═══ Highway to be built 'with guaranteed free and peaceful passage by Egypt and Jordan'.

▨ Area of limited Egyptian forces (50 Kilometres)

▬ Area of limited Israeli forces (3 Kilometres)

⬚ Approximate area to be policed by only United Nations forces and 'civil police equipped with light weapons'. (20 to 40 Kilometres)

◁ Western boundary of the first phase of Israeli troop withdrawals, to be taken up 'between three months and nine months' after the signing of 'a peace Treaty between Egypt and Israel.

© Martin Gilbert 1978

110

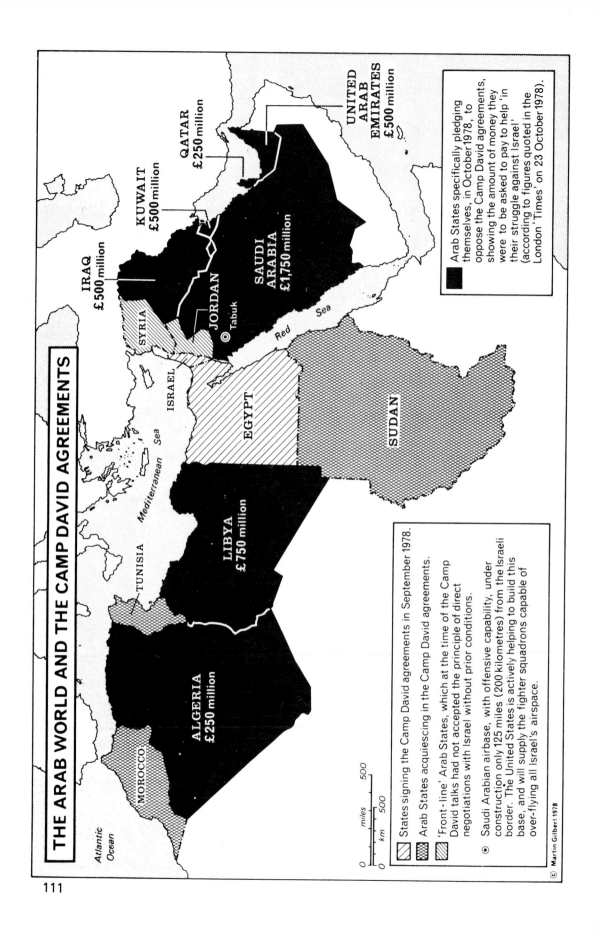

THE ARAB WORLD AND THE CAMP DAVID AGREEMENTS

UNITED ARAB EMIRATES £500 million

QATAR £250 million

KUWAIT £500 million

SAUDI ARABIA £1,750 million

IRAQ £500 million

JORDAN

SYRIA

⊙ Tabuk

Red Sea

ISRAEL

EGYPT

SUDAN

Mediterranean Sea

TUNISIA

LIBYA £750 million

ALGERIA £250 million

MOROCCO

Atlantic Ocean

Arab States specifically pledging themselves, in October 1978, to oppose the Camp David agreements, showing the amount of money they were to be asked to pay to help 'in their struggle against Israel' (according to figures quoted in the London 'Times' on 23 October 1978).

States signing the Camp David agreements in September 1978.

Arab States acquiescing in the Camp David agreements.

'Front-line' Arab States, which at the time of the Camp David talks had not accepted the principle of direct negotiations with Israel without prior conditions.

⊙ Saudi Arabian airbase, with offensive capability, under construction only 125 miles (200 kilometres) from the Israeli border. The United States is actively helping to build this base, and will supply the fighter squadrons capable of over-flying all Israel's airspace.

miles 500

km 500

© Martin Gilbert 1978

111

ISLAMIC FUNDAMENTALISM SINCE 1979

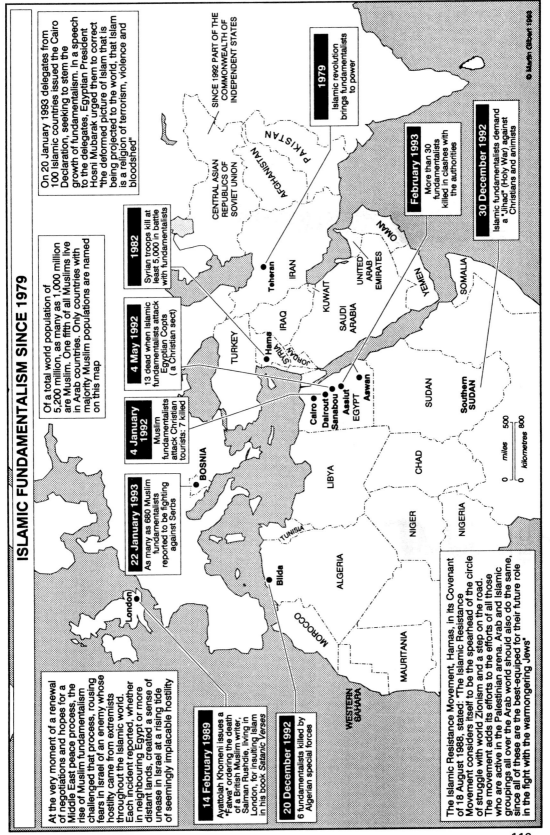

At the very moment of a renewal of negotiations and hopes for a Middle East peace process, the rise of Muslim fundamentalism challenged that process, rousing fears in Israel of an enemy whose hostility came from extremists throughout the Islamic world. Each incident reported, whether in neighbouring Egypt or more distant lands, created a sense of unease in Israel at a rising tide of seemingly implacable hostility

Of a total world population of 5,200 million, as many as 1,000 million are Muslim. One fifth of all Muslims live in Arab countries. Only countries with majority Muslim populations are named on this map

On 20 January 1993 delegates from 100 Islamic countries issued the Cairo Declaration, seeking to stem the growth of fundamentalism. In a speech to the delegates, Egyptian President Hosni Mubarak urged them to correct "the deformed picture of Islam that is being projected to the world, that Islam is a religion of terrorism, violence and bloodshed"

1979
Islamic revolution brings fundamentalists to power

February 1993
More than 30 fundamentalists killed in clashes with the authorities

30 December 1992
Islamic fundamentalists demand a "Jihad" (Holy War) against Christians and animists

1982
Syrian troops kill at least 5,000 in battle with fundamentalists

4 May 1992
13 dead when Islamic fundamentalists attack Egyptian Copts (a Christian sect)

4 January 1992
Muslim fundamentalists attack Christian tourists: 7 killed

22 January 1993
As many as 680 Muslim fundamentalists reported to be fighting against Serbs

14 February 1989
Ayatollah Khomeni issues a "Fatwa' ordering the death of a British Muslim writer, Salman Rushdie, living in London, for insulting Islam in his book *Satanic Verses*

20 December 1992
6 fundamentalists killed by Algerian special forces

The Islamic Resistance Movement, Hamas, in its Covenant of 18 August 1988, stated: 'The Islamic Resistance Movement considers itself to be the spearhead of the circle of struggle with world Zionism and a step on the road. The movement adds its efforts to the efforts of all those who are active in the Palestinian arena. Arab and Islamic groupings all over the Arab world should also do the same, since all of these are the best-equiped for their future role in the fight with the warmongering Jews"

SINCE 1992 PART OF THE COMMONWEALTH OF INDEPENDENT STATES

CENTRAL ASIAN REPUBLICS OF SOVIET UNION

PAKISTAN

AFGHANISTAN

Teheran •

IRAN

OMAN

KUWAIT

UNITED ARAB EMIRATES

SAUDI ARABIA

YEMEN

SOMALIA

IRAQ

TURKEY

Hama •

SYRIA

JORDAN

Cairo •
Dairout •
Sanabou •
Assiut •
Aswan •
EGYPT

SUDAN

Southern SUDAN

BOSNIA

LIBYA

CHAD

NIGER

NIGERIA

TUNISIA

ALGERIA

Blida •

MOROCCO

MAURITANIA

WESTERN SAHARA

London •

0   500 miles
0   800 kilometres

© Martin Gilbert 1993

112

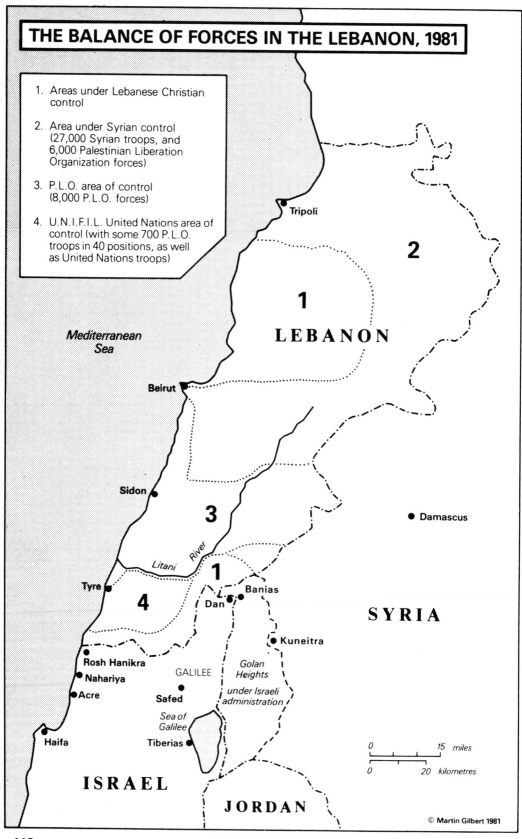

THE BALANCE OF FORCES IN THE LEBANON, 1981

1. Areas under Lebanese Christian control

2. Area under Syrian control (27,000 Syrian troops, and 6,000 Palestinian Liberation Organization forces)

3. P.L.O. area of control (8,000 P.L.O. forces)

4. U.N.I.F.I.L. United Nations area of control (with some 700 P.L.O. troops in 40 positions, as well as United Nations troops)

Mediterranean Sea

Tripoli

2

1

L E B A N O N

Beirut

Sidon

3

Litani River

Damascus

1

Tyre

Banias

Dan

4

S Y R I A

Kuneitra

Rosh Hanikra

Nahariya

GALILEE

Golan Heights

Acre

Safed

under Israeli administration

Sea of Galilee

Haifa

Tiberias

I S R A E L

J O R D A N

0   15 *miles*

0   20 *kilometres*

© Martin Gilbert 1981

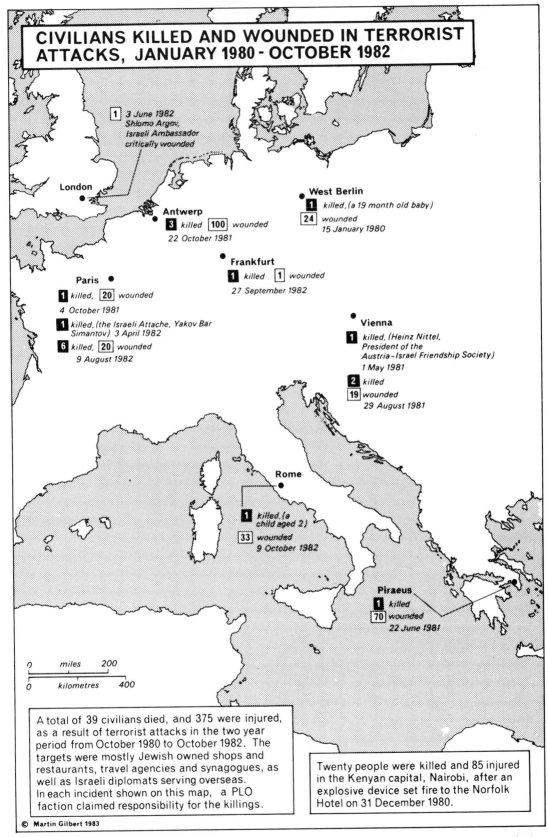

CIVILIANS KILLED AND WOUNDED IN TERRORIST ATTACKS, JANUARY 1980 - OCTOBER 1982

1 3 June 1982
Shlomo Argov,
Israeli Ambassador
critically wounded

London

West Berlin
1 killed, (a 19 month old baby)
24 wounded
15 January 1980

Antwerp
3 killed **100** wounded
22 October 1981

Frankfurt
1 killed **1** wounded
27 September 1982

Paris
1 killed, **20** wounded
4 October 1981
1 killed, (the Israeli Attache, Yakov Bar
Simantov) 3 April 1982
6 killed, **20** wounded
9 August 1982

Vienna
1 killed, (Heinz Nittel,
President of the
Austria-Israel Friendship Society)
1 May 1981
2 killed
19 wounded
29 August 1981

Rome
1 killed, (a
child aged 2)
33 wounded
9 October 1982

Piraeus
1 killed
70 wounded
22 June 1981

miles 200

0

kilometres 400

0

A total of 39 civilians died, and 375 were injured,
as a result of terrorist attacks in the two year
period from October 1980 to October 1982. The
targets were mostly Jewish owned shops and
restaurants, travel agencies and synagogues, as
well as Israeli diplomats serving overseas.
In each incident shown on this map, a PLO
faction claimed responsibility for the killings.

Twenty people were killed and 85 injured
in the Kenyan capital, Nairobi, after an
explosive device set fire to the Norfolk
Hotel on 31 December 1980.

© Martin Gilbert 1983

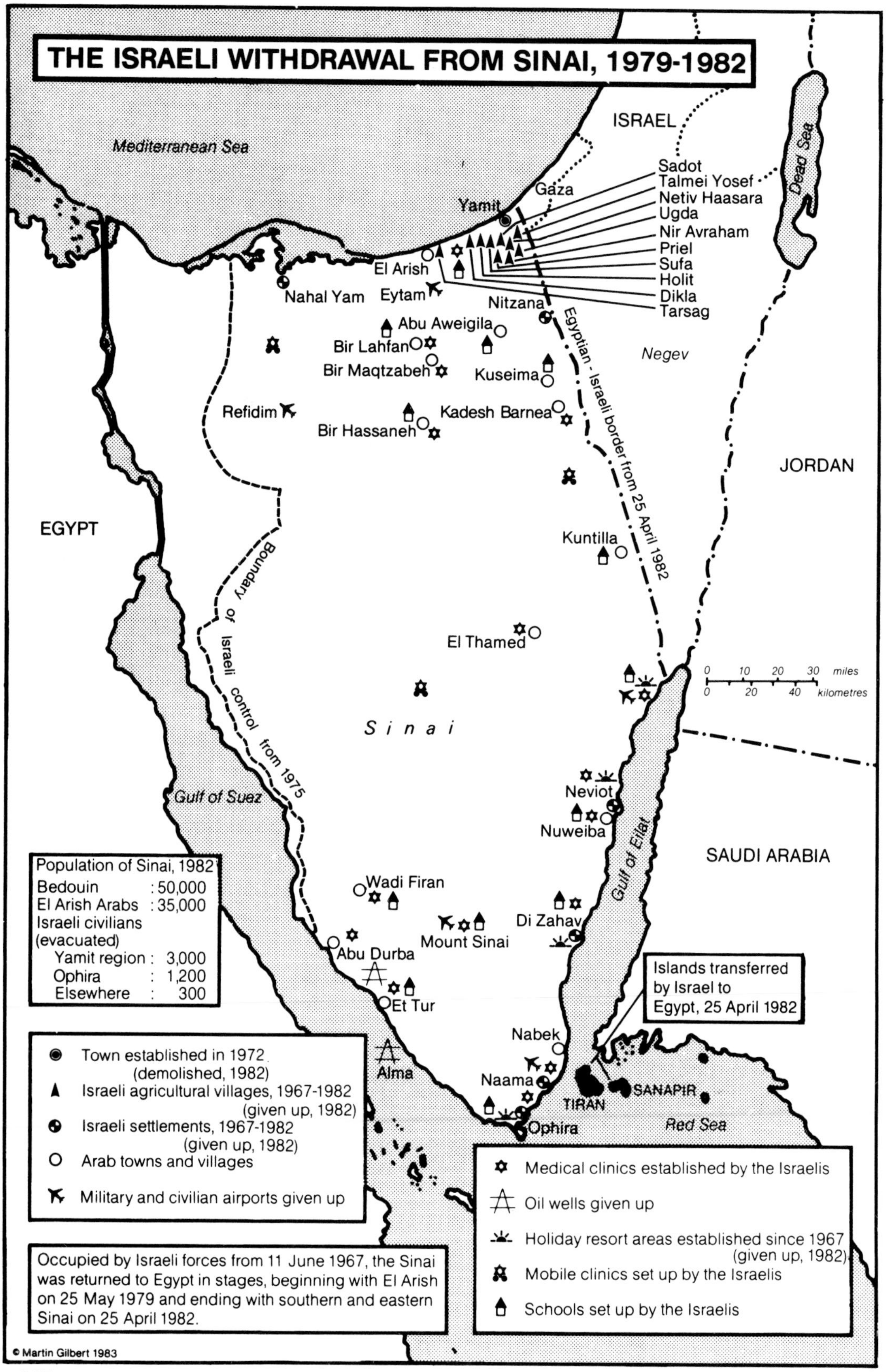
THE ISRAELI WITHDRAWAL FROM SINAI, 1979-1982
Mediterranean Sea
ISRAEL
Dead Sea
Sadot
Talmei Yosef
Netiv Haasara
Ugda
Nir Avraham
Priel
Sufa
Holit
Dikla
Tarsag
Gaza
Yamit
El Arish
Nahal Yam
Eytam
Nitzana
Abu Aweigila
Bir Lahfan
Bir Maqtzabeh
Kuseima
Negev
Refidim
Kadesh Barnea
Bir Hassaneh
Egyptian - Israeli border from 25 April 1982
JORDAN
EGYPT
Kuntilla
Boundary of Israeli control from 1975
El Thamed
0 10 20 30 miles
0 20 40 kilometres
Sinai
Gulf of Suez
Neviot
Nuweiba
Gulf of Eilat
SAUDI ARABIA
Population of Sinai, 1982
Bedouin : 50,000
El Arish Arabs : 35,000
Israeli civilians (evacuated)
Yamit region : 3,000
Ophira : 1,200
Elsewhere : 300
Wadi Firan
Mount Sinai
Di Zahav
Abu Durba
Et Tur
Islands transferred by Israel to Egypt, 25 April 1982
Nabek
Alma
Naama
TIRAN
SANAPIR
Ophira
Red Sea
Town established in 1972 (demolished, 1982)
Israeli agricultural villages, 1967-1982 (given up, 1982)
Israeli settlements, 1967-1982 (given up, 1982)
Arab towns and villages
Military and civilian airports given up
Medical clinics established by the Israelis
Oil wells given up
Holiday resort areas established since 1967 (given up, 1982)
Mobile clinics set up by the Israelis
Schools set up by the Israelis
Occupied by Israeli forces from 11 June 1967, the Sinai was returned to Egypt in stages, beginning with El Arish on 25 May 1979 and ending with southern and eastern Sinai on 25 April 1982.
© Martin Gilbert 1983

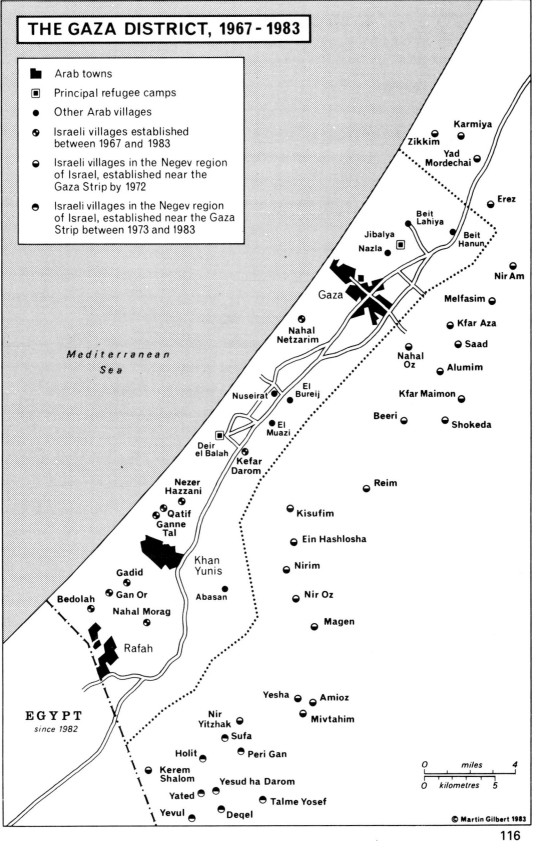

THE GAZA DISTRICT, 1967 - 1983

Arab towns

Principal refugee camps

Other Arab villages

Israeli villages established between 1967 and 1983

Israeli villages in the Negev region of Israel, established near the Gaza Strip by 1972

Israeli villages in the Negev region of Israel, established near the Gaza Strip between 1973 and 1983

Karmiya

Zikkim

Yad Mordechai

Erez

Beit Lahiya

Jibalya

Nazla

Beit Hanun

Nir Am

Gaza

Melfasim

Kfar Aza

Nahal Netzarim

Saad

Nahal Oz

Alumim

Kfar Maimon

Mediterranean Sea

Nuseirat

El Bureij

Beeri

Shokeda

El Muazi

Deir el Balah

Kefar Darom

Reim

Nezer Hazzani

Kisufim

Qatif

Ganne Tal

Ein Hashlosha

Khan Yunis

Nirim

Gadid

Gan Or

Abasan

Nir Oz

Bedolah

Nahal Morag

Magen

Rafah

Yesha

Amioz

Mivtahim

EGYPT
since 1982

Nir Yitzhak

Sufa

Holit

Peri Gan

Kerem Shalom

Yesud ha Darom

Yated

Talme Yosef

Yevul

Deqel

| 0 | | miles | | 4 |

| 0 | | kilometres | | 5 |

© Martin Gilbert 1983

116

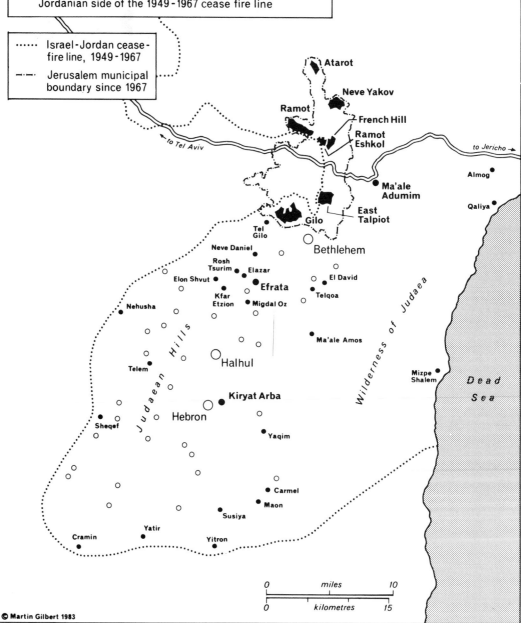

ISRAELI SETTLEMENTS IN JUDAEA, 1967 – 1983

JERUSALEM POPULATION, 1982	
Jews	305,000
Muslim Arabs	109,000
Christian Arabs	11,000
Total population	425,000

○   Principal Arab towns

o   Arab villages

●   Israeli settlements set up between 1967 and 1983, intended for up to 1,500 families each.

•   Israeli settlements set up between 1967 and 1983, intended for up to 300 families each.

◄   Jerusalem suburbs built by Israel on the former Jordanian side of the 1949 - 1967 cease fire line

·········   Israel - Jordan cease-fire line, 1949 - 1967

–·–·–   Jerusalem municipal boundary since 1967

Atarot

Neve Yakov

Ramot

French Hill

Ramot Eshkol

to Tel Aviv

to Jericho

Almog

Ma'ale Adumim

East Talpiot

Qaliya

Gilo

Tel Gilo

Bethlehem

Neve Daniel

Rosh Tsurim

Elazar

El David

Elon Shvut

Efrata

Kfar Etzion

Telqoa

Migdal Oz

Nehusha

Ma'ale Amos

Wilderness of Judaea

Judaean Hills

Halhul

Telem

Mizpe Shalem

Dead Sea

Kiryat Arba

Hebron

Sheqef

Yaqim

Carmel

Maon

Susiya

Cramin

Yatir

Yitron

0   miles   10

0   kilometres   15

© Martin Gilbert 1983

117

THE WEST BANK: COMPARATIVE SIZES

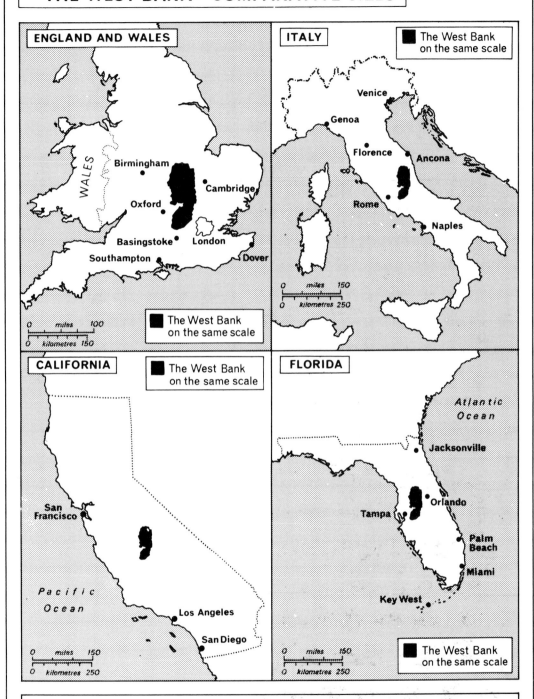

ENGLAND AND WALES

WALES
Birmingham
Cambridge
Oxford
Basingstoke
London
Southampton
Dover

■ The West Bank on the same scale

0 miles 100
0 kilometres 150

ITALY

■ The West Bank on the same scale

Venice
Genoa
Florence
Ancona
Rome
Naples

0 miles 150
0 kilometres 250

CALIFORNIA

■ The West Bank on the same scale

San Francisco
Los Angeles
San Diego
Pacific Ocean

0 miles 150
0 kilometres 250

FLORIDA

Atlantic Ocean
Jacksonville
Orlando
Tampa
Palm Beach
Miami
Key West

■ The West Bank on the same scale

0 miles 150
0 kilometres 250

A total of 1,186,000 Arabs were living on the West Bank (721,700) and in the Gaza Strip (464,300) in 1982. The number of Israeli settlers was approximately 20,000. According to Israeli Government plans, it was hoped to raise this figure to 100,000 by the end of the decade, but the main Israeli Opposition parties were opposed to any such expansion, and many of the new settlements shown on the Judaea and Samaria maps had as few as 26 families in July 1983.

© Martin Gilbert 1983

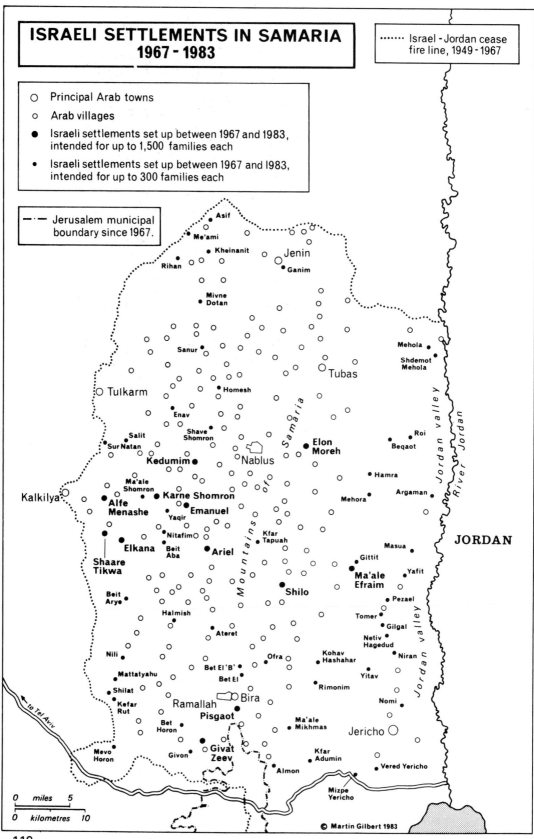

ISRAELI SETTLEMENTS IN SAMARIA 1967-1983

······· Israel - Jordan cease fire line, 1949 - 1967

○ Principal Arab towns

○ Arab villages

● Israeli settlements set up between 1967 and 1983, intended for up to 1,500 families each

• Israeli settlements set up between 1967 and 1983, intended for up to 300 families each

—··— Jerusalem municipal boundary since 1967.

Asif

Me'ami

Kheinanit

Jenin

Rihan

Ganim

Mivne Dotan

Sanur

Mehola

Shdemot Mehola

Tubas

Tulkarm

Homesh

Samaria

Enav

Roi

Salit

Shave Shomron

Beqaot

Sur Natan

Elon Moreh

Kedumim

Nablus

Ma'ale Shomron

Hamra

Kalkilya

Karne Shomron

Mehora

Argaman

Alfe Menashe

Emanuel

Yaqir

JORDAN

Nitafim

Kfar Tapuah

Masua

Elkana

Beit Aba

Ariel

Gittit

Yafit

Shaare Tikwa

Ma'ale Efraim

Mountains

Shilo

Pezael

Beit Arye

Tomer

Gilgal

Halmish

Netiv Hagedud

Ateret

Kohav Hashahar

Niran

Nili

Ofra

Yitav

Bet El 'B'

Mattatyahu

Bet El

Rimonim

Shilat

Nomi

Kefar Rut

Ramallah

Bira

Pisgaot

Bet Horon

Ma'ale Mikhmas

Jericho

Mevo Horon

Givon

Givat Zeev

Kfar Adumin

Almon

Vered Yericho

Jordan valley

River Jordan

Mizpe Yericho

to Tel Aviv

0 miles 5

0 kilometres 10

© Martin Gilbert 1983

119

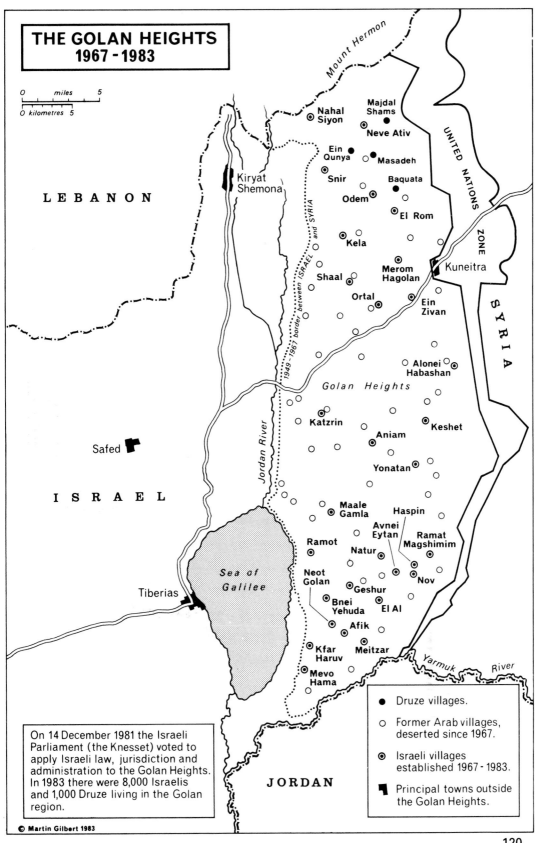

THE GOLAN HEIGHTS
1967-1983

O miles 5
O kilometres 5

LEBANON

Kiryat Shemona

1949 - 1967 border between ISRAEL and SYRIA

Mount Hermon

Nahal Siyon

Majdal Shams

Neve Ativ

Ein Qunya
Masadeh

Snir
Baquata

Odem
El Rom

Kela

Shaal
Merom Hagolan
Kuneitra

Ortal
Ein Zivan

Golan Heights

Alonei Habashan

ISRAEL

Safed

Katzrin
Keshet

Aniam

Yonatan

Jordan River

Maale Gamla
Haspin

Avnei Eytan
Ramat Magshimim

Ramot
Natur
Nov

Neot Golan
Geshur

Sea of Galilee

Bnei Yehuda
El Al

Tiberias

Afik

Kfar Haruv
Meitzar

Mevo Hama

Yarmuk River

UNITED NATIONS ZONE

SYRIA

● Druze villages.

○ Former Arab villages, deserted since 1967.

◉ Israeli villages established 1967 - 1983.

◤ Principal towns outside the Golan Heights.

On 14 December 1981 the Israeli Parliament (the Knesset) voted to apply Israeli law, jurisdiction and administration to the Golan Heights. In 1983 there were 8,000 Israelis and 1,000 Druze living in the Golan region.

© Martin Gilbert 1983

JORDAN

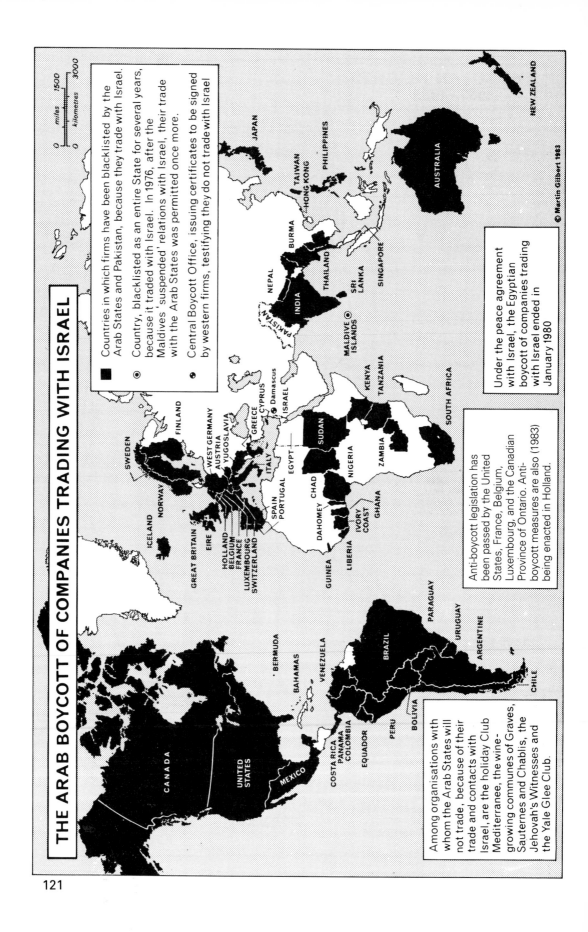

THE ARAB BOYCOTT OF COMPANIES TRADING WITH ISRAEL

Among organisations with whom the Arab States will not trade, because of their trade and contacts with Israel, are the holiday Club Mediterranee, the wine-growing communes of Graves, Sauternes and Chablis, the Jehovah's Witnesses and the Yale Glee Club.

Anti-boycott legislation has been passed by the United States, France, Belgium, Luxembourg, and the Canadian Province of Ontario. Anti-boycott measures are also (1983) being enacted in Holland.

Under the peace agreement with Israel, the Egyptian boycott of companies trading with Israel ended in January 1980

■ Countries in which firms have been blacklisted by the Arab States and Pakistan, because they trade with Israel.

◉ Country, blacklisted as an entire State for several years, because it traded with Israel. In 1976, after the Maldives 'suspended' relations with Israel, their trade with the Arab States was permitted once more.

● Central Boycott Office, issuing certificates to be signed by western firms, testifying they do not trade with Israel

© Martin Gilbert 1983

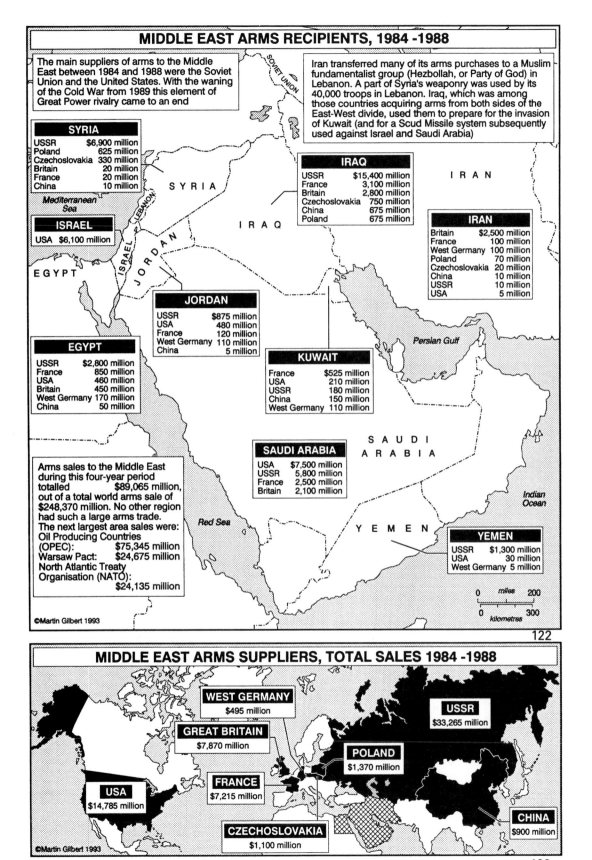

MIDDLE EAST ARMS RECIPIENTS, 1984-1988

The main suppliers of arms to the Middle East between 1984 and 1988 were the Soviet Union and the United States. With the waning of the Cold War from 1989 this element of Great Power rivalry came to an end

Iran transferred many of its arms purchases to a Muslim fundamentalist group (Hezbollah, or Party of God) in Lebanon. A part of Syria's weaponry was used by its 40,000 troops in Lebanon. Iraq, which was among those countries acquiring arms from both sides of the East-West divide, used them to prepare for the invasion of Kuwait (and for a Scud Missile system subsequently used against Israel and Saudi Arabia)

SYRIA

USSR	$6,900 million
Poland	625 million
Czechoslovakia	330 million
Britain	20 million
France	20 million
China	10 million

ISRAEL

USA	$6,100 million

IRAQ

USSR	$15,400 million
France	3,100 million
Britain	2,800 million
Czechoslovakia	750 million
China	675 million
Poland	675 million

IRAN

Britain	$2,500 million
France	100 million
West Germany	100 million
Poland	70 million
Czechoslovakia	20 million
China	10 million
USSR	10 million
USA	5 million

JORDAN

USSR	$875 million
USA	480 million
France	120 million
West Germany	110 million
China	5 million

EGYPT

USSR	$2,800 million
France	850 million
USA	460 million
Britain	450 million
West Germany	170 million
China	50 million

KUWAIT

France	$525 million
USA	210 million
USSR	180 million
China	150 million
West Germany	110 million

SAUDI ARABIA

USA	$7,500 million
USSR	5,800 million
France	2,500 million
Britain	2,100 million

Arms sales to the Middle East during this four-year period totalled $89,065 million, out of a total world arms sale of $248,370 million. No other region had such a large arms trade. The next largest area sales were:
Oil Producing Countries (OPEC): $75,345 million
Warsaw Pact: $24,675 million
North Atlantic Treaty Organisation (NATO): $24,135 million

YEMEN

USSR	$1,300 million
USA	30 million
West Germany	5 million

0 — miles — 200
0 — kilometres — 300

©Martin Gilbert 1993

122

MIDDLE EAST ARMS SUPPLIERS, TOTAL SALES 1984-1988

WEST GERMANY
$495 million

USSR
$33,265 million

GREAT BRITAIN
$7,870 million

POLAND
$1,370 million

USA
$14,785 million

FRANCE
$7,215 million

CHINA
$900 million

CZECHOSLOVAKIA
$1,100 million

©Martin Gilbert 1993

123

THE WAR IN LEBANON, 6 JUNE 1982 – 21 AUGUST 1982

Following the attempted assassination of the Israeli Ambassador to London, Shlomo Argov, on 3 June 1982, and renewed PLO shelling of northern Israel, Israeli forces launched 'Operation Peace for Galilee' on 6 June 1982. Fighting continued until 21 August 1982. Following the start of direct Lebanese–Israeli talks on 28 December 1982, agreement was signed on 17 May 1983 for the withdrawal of all foreign forces from Lebanon. The Syrian Government, however, rejected the call for the withdrawal of Syrian troops.

ISRAELI ESTIMATE
OF WAR DEATHS

Israeli forces	368
Syrian forces	600
P.L.O. forces	3,000
Lebanese civilians	
(outside Beirut)	460
(in Beirut)	unknown

According to Arab sources, Arab civilian deaths were in excess of ten thousand

Beirut

Zahle

Legend:
- ▮ Dates of Israeli advance
- ← Main direction of Israeli advance
- ▲▲▲ Israeli front line by 11 June 1982

Beirut – Damascus highway

Baabda

Aley

Shtoura

Bekaa Valley

8 JUNE

8 JUNE

11 JUNE

SYRIA

Damour

7 JUNE

6 JUNE
Sea-borne landing

Awali River

10 JUNE

11 JUNE

Sidon

Jezzine

7 JUNE

10 JUNE

Mediterranean Sea

Rachaiya

7 JUNE

8 JUNE **8 JUNE**

Zaharani River

Hasbaiya

Hammadiye

Marjayoun

Mount Hermon

Beaufort

6 JUNE

6 JUNE

Litani River

6 JUNE

SYRIA

Tyre

Rachidye

LEBANON

Kiryat Shmona

Golan Heights

6 JUNE

ISRAEL

Kuneitra

Bint Jubail

0		5		10	miles
0	5	10	15		kilometres

© Martin Gilbert 1983

124

SABRA AND CHATILA, 16 - 17 SEPTEMBER 1982

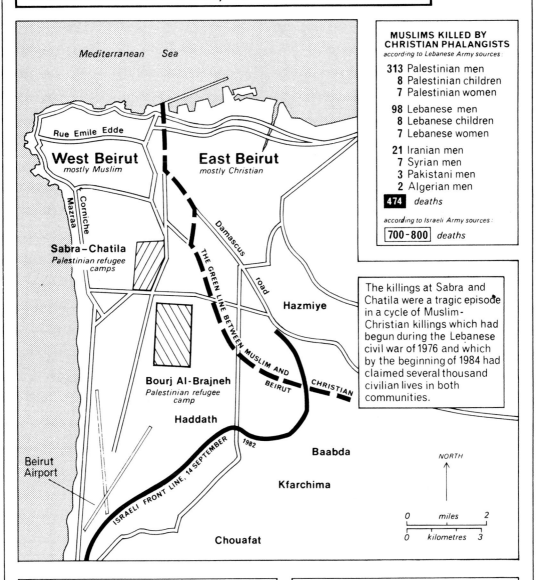

MUSLIMS KILLED BY CHRISTIAN PHALANGISTS
according to Lebanese Army sources:

313 Palestinian men
8 Palestinian children
7 Palestinian women

98 Lebanese men
8 Lebanese children
7 Lebanese women

21 Iranian men
7 Syrian men
3 Pakistani men
2 Algerian men

474 *deaths*

according to Israeli Army sources:

700-800 *deaths*

The killings at Sabra and Chatila were a tragic episode in a cycle of Muslim-Christian killings which had begun during the Lebanese civil war of 1976 and which by the beginning of 1984 had claimed several thousand civilian lives in both communities.

Mediterranean Sea

Rue Emile Edde

West Beirut *mostly Muslim*

East Beirut *mostly Christian*

Mazraa
Corniche

Damascus road

THE GREEN LINE BETWEEN MUSLIM AND BEIRUT CHRISTIAN

Sabra - Chatila *Palestinian refugee camps*

Hazmiye

Bourj Al-Brajneh *Palestinian refugee camp*

Haddath

ISRAELI FRONT LINE, 14 SEPTEMBER 1982

Beirut Airport

Baabda

Kfarchima

NORTH ↑

0 miles 2
0 kilometres 3

Chouafat

---

14 September: Bashir Gemayel, President Elect of Lebanon, killed in a bomb explosion

15 September: Israeli forces occupy West Beirut 'in order to protect the Muslims from the vengeance of the Phalangists' **(Israeli PM, Menachem Begin)**

16 September: 'Christian Phalangist forces begin sweeping camps in West Beirut. A massacre of men, women and children took place in the Sabra and Chatila refugee camps'. **(The Times, 3 January 1983)**

23 September: Amin Gemayel elected President

28 September: Israeli forces leave West Beirut

An Israeli judicial enquiry, the Kahan Report, concluded: 'No intention existed on the part of any Israeli element to harm the non-combatant population in the camps'. But the Director of Israeli Military Intelligence was criticized for 'closing his eyes and blocking his ears' ; the Israeli Minister of Defence (Ariel Sharon) because humanitarian obligations 'did not concern him in the least'; and the Israeli Prime Minister (Menachem Begin) for unjustifiable 'indifference', and 'for not having evinced during or after the Cabinet session any interest in the Phalangists' actions in the camps'. **(Kahan Report, 7 February 1983)**

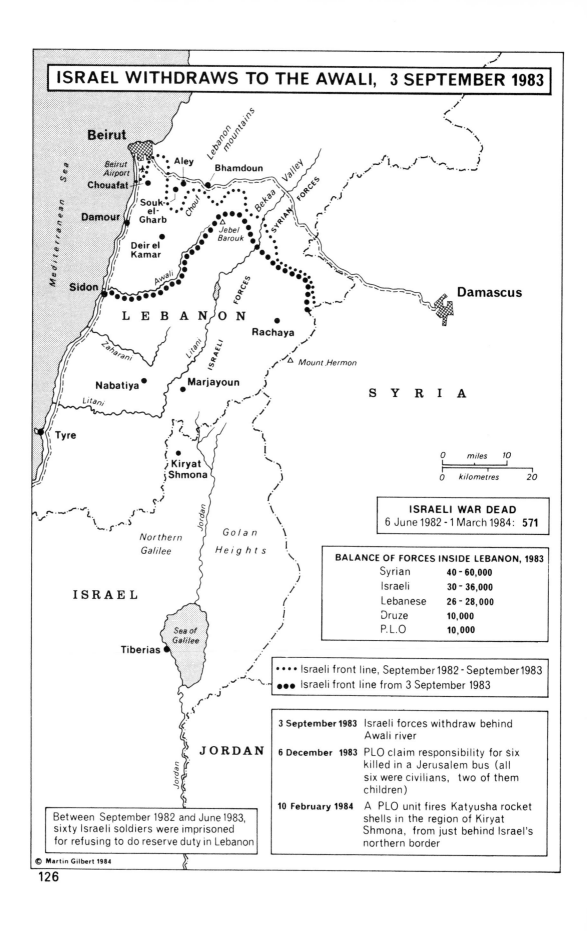

ISRAEL WITHDRAWS TO THE AWALI, 3 SEPTEMBER 1983

Beirut

Beirut Airport

Aley

Bhamdoun

Lebanon mountains

Chouafat

Souk-el-Gharb

Chouf

Bekaa Valley

SYRIAN FORCES

Damour

Deir el Kamar

Jebel Barouk

Mediterranean Sea

Awali

Sidon

L E B A N O N

Rachaya

Zaharani

Litani

ISRAELI FORCES

Mount Hermon

S Y R I A

Damascus

Nabatiya

Marjayoun

Litani

Tyre

Northern Galilee

Golan Heights

Jordan

Kiryat Shmona

I S R A E L

Sea of Galilee

Tiberias

JORDAN

Jordan

0   miles   10

0   kilometres   20

ISRAELI WAR DEAD
6 June 1982 - 1 March 1984 : **571**

BALANCE OF FORCES INSIDE LEBANON, 1983

Syrian	**40 - 60,000**
Israeli	**30 - 36,000**
Lebanese	**26 - 28,000**
Druze	**10,000**
P.L.O	**10,000**

•••• Israeli front line, September 1982 - September 1983

●●● Israeli front line from 3 September 1983

---

3 September 1983	Israeli forces withdraw behind Awali river
6 December 1983	PLO claim responsibility for six killed in a Jerusalem bus (all six were civilians, two of them children)
10 February 1984	A PLO unit fires Katyusha rocket shells in the region of Kiryat Shmona, from just behind Israel's northern border

Between September 1982 and June 1983, sixty Israeli soldiers were imprisoned for refusing to do reserve duty in Lebanon

© Martin Gilbert 1984

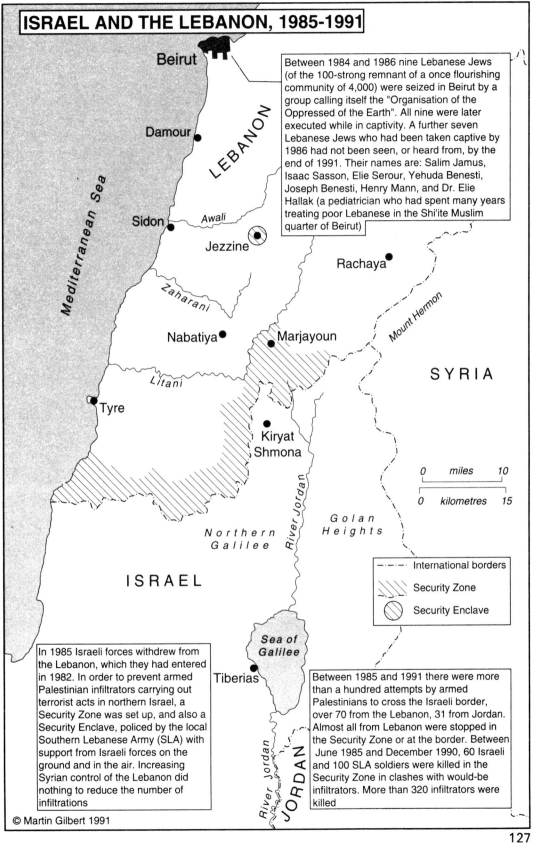

ISRAEL AND THE LEBANON, 1985-1991

Beirut

Damour

LEBANON

Mediterranean Sea

Between 1984 and 1986 nine Lebanese Jews (of the 100-strong remnant of a once flourishing community of 4,000) were seized in Beirut by a group calling itself the "Organisation of the Oppressed of the Earth". All nine were later executed while in captivity. A further seven Lebanese Jews who had been taken captive by 1986 had not been seen, or heard from, by the end of 1991. Their names are: Salim Jamus, Isaac Sasson, Elie Serour, Yehuda Benesti, Joseph Benesti, Henry Mann, and Dr. Elie Hallak (a pediatrician who had spent many years treating poor Lebanese in the Shi'ite Muslim quarter of Beirut)

Sidon   *Awali*

Jezzine

Rachaya

Zaharani

Nabatiya   Marjayoun

Mount Hermon

S Y R I A

Litani

Tyre

Kiryat
Shmona

River Jordan

0   miles   10

0   kilometres   15

*G o l a n
H e i g h t s*

*N o r t h e r n
G a l i l e e*

I S R A E L

---·--- International borders

////// Security Zone

◯ Security Enclave

*Sea of
Galilee*

In 1985 Israeli forces withdrew from the Lebanon, which they had entered in 1982. In order to prevent armed Palestinian infiltrators carrying out terrorist acts in northern Israel, a Security Zone was set up, and also a Security Enclave, policed by the local Southern Lebanese Army (SLA) with support from Israeli forces on the ground and in the air. Increasing Syrian control of the Lebanon did nothing to reduce the number of infiltrations

Tiberias

Between 1985 and 1991 there were more than a hundred attempts by armed Palestinians to cross the Israeli border, over 70 from the Lebanon, 31 from Jordan. Almost all from Lebanon were stopped in the Security Zone or at the border. Between June 1985 and December 1990, 60 Israeli and 100 SLA soldiers were killed in the Security Zone in clashes with would-be infiltrators. More than 320 infiltrators were killed

River Jordan

J O R D A N

© Martin Gilbert 1991

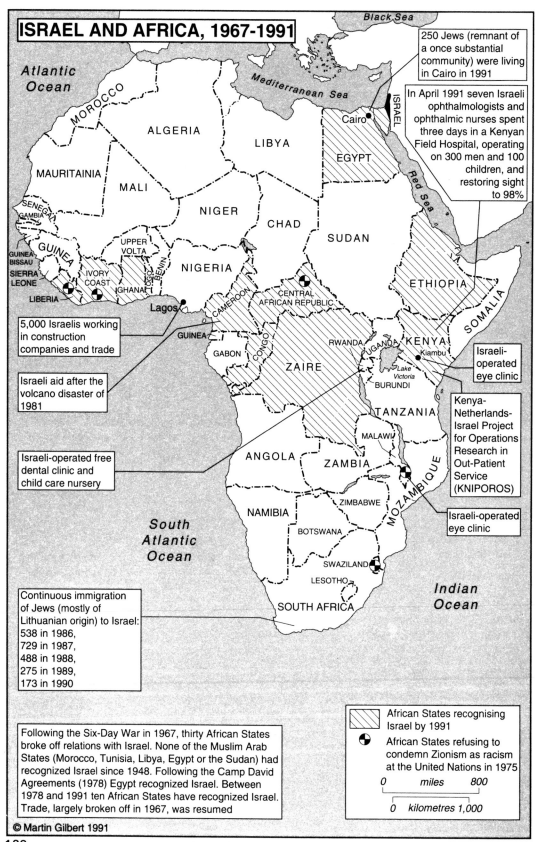

ISRAEL AND AFRICA, 1967-1991

Atlantic Ocean

Black Sea

Mediterranean Sea

250 Jews (remnant of a once substantial community) were living in Cairo in 1991

In April 1991 seven Israeli ophthalmologists and ophthalmic nurses spent three days in a Kenyan Field Hospital, operating on 300 men and 100 children, and restoring sight to 98%

MOROCCO

ALGERIA

LIBYA

EGYPT

ISRAEL

Cairo

Red Sea

MAURITAINIA

MALI

NIGER

CHAD

SUDAN

SENEGAL

GAMBIA

GUINEA

UPPER VOLTA

NIGERIA

CAMEROON

CENTRAL AFRICAN REPUBLIC

ETHIOPIA

SOMALIA

GUINEA BISSAU

SIERRA LEONE

IVORY COAST

TOGO

BENIN

GHANA

LIBERIA

Lagos

5,000 Israelis working in construction companies and trade

GUINEA

GABON

CONGO

ZAIRE

RWANDA

UGANDA

KENYA

Kiambu

Israeli-operated eye clinic

Lake Victoria

BURUNDI

Kenya-Netherlands-Israel Project for Operations Research in Out-Patient Service (KNIPOROS)

Israeli aid after the volcano disaster of 1981

TANZANIA

MALAWI

Israeli-operated free dental clinic and child care nursery

ANGOLA

ZAMBIA

ZIMBABWE

MOZAMBIQUE

Israeli-operated eye clinic

NAMIBIA

BOTSWANA

South Atlantic Ocean

SWAZILAND

LESOTHO

SOUTH AFRICA

Indian Ocean

Continuous immigration of Jews (mostly of Lithuanian origin) to Israel:
538 in 1986,
729 in 1987,
488 in 1988,
275 in 1989,
173 in 1990

African States recognising Israel by 1991

African States refusing to condemn Zionism as racism at the United Nations in 1975

0   miles   800

0   kilometres 1,000

Following the Six-Day War in 1967, thirty African States broke off relations with Israel. None of the Muslim Arab States (Morocco, Tunisia, Libya, Egypt or the Sudan) had recognized Israel since 1948. Following the Camp David Agreements (1978) Egypt recognized Israel. Between 1978 and 1991 ten African States have recognized Israel. Trade, largely broken off in 1967, was resumed

© Martin Gilbert 1991

128

THE WEST BANK, 1967-1991

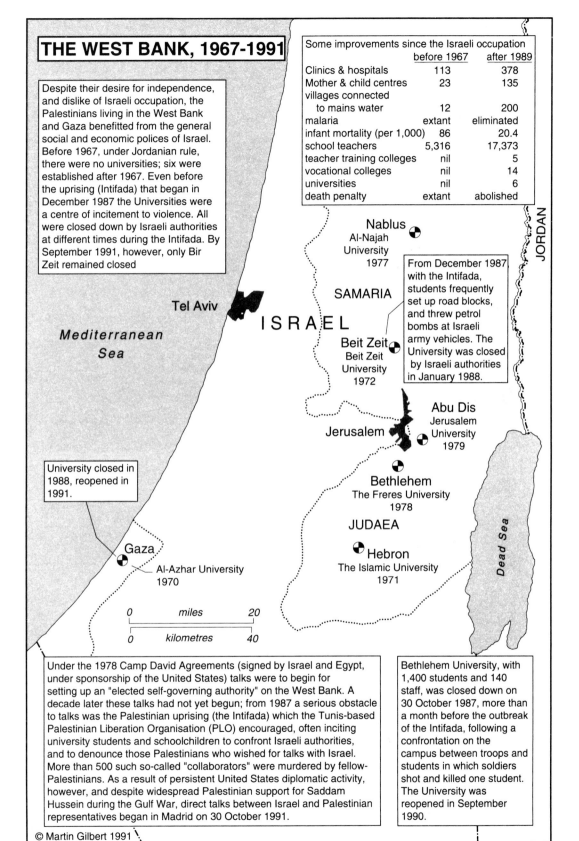

Despite their desire for independence, and dislike of Israeli occupation, the Palestinians living in the West Bank and Gaza benefitted from the general social and economic polices of Israel. Before 1967, under Jordanian rule, there were no universities; six were established after 1967. Even before the uprising (Intifada) that began in December 1987 the Universities were a centre of incitement to violence. All were closed down by Israeli authorities at different times during the Intifada. By September 1991, however, only Bir Zeit remained closed

Some improvements since the Israeli occupation

	before 1967	after 1989
Clinics & hospitals	113	378
Mother & child centres	23	135
villages connected to mains water	12	200
malaria	extant	eliminated
infant mortality (per 1,000)	86	20.4
school teachers	5,316	17,373
teacher training colleges	nil	5
vocational colleges	nil	14
universities	nil	6
death penalty	extant	abolished

JORDAN

Nablus
Al-Najah
University
1977

SAMARIA

Tel Aviv

I S R A E L

*Mediterranean
Sea*

From December 1987, with the Intifada, students frequently set up road blocks, and threw petrol bombs at Israeli army vehicles. The University was closed by Israeli authorities in January 1988.

Beit Zeit
Beit Zeit
University
1972

Abu Dis
Jerusalem
University
1979

Jerusalem

University closed in 1988, reopened in 1991.

Bethlehem
The Freres University
1978

JUDAEA

Dead Sea

Gaza
Al-Azhar University
1970

Hebron
The Islamic University
1971

0	miles	20
0	kilometres	40

Under the 1978 Camp David Agreements (signed by Israel and Egypt, under sponsorship of the United States) talks were to begin for setting up an "elected self-governing authority" on the West Bank. A decade later these talks had not yet begun; from 1987 a serious obstacle to talks was the Palestinian uprising (the Intifada) which the Tunis-based Palestinian Liberation Organisation (PLO) encouraged, often inciting university students and schoolchildren to confront Israeli authorities, and to denounce those Palestinians who wished for talks with Israel. More than 500 such so-called "collaborators" were murdered by fellow-Palestinians. As a result of persistent United States diplomatic activity, however, and despite widespread Palestinian support for Saddam Hussein during the Gulf War, direct talks between Israel and Palestinian representatives began in Madrid on 30 October 1991.

Bethlehem University, with 1,400 students and 140 staff, was closed down on 30 October 1987, more than a month before the outbreak of the Intifada, following a confrontation on the campus between troops and students in which soldiers shot and killed one student. The University was reopened in September 1990.

© Martin Gilbert 1991

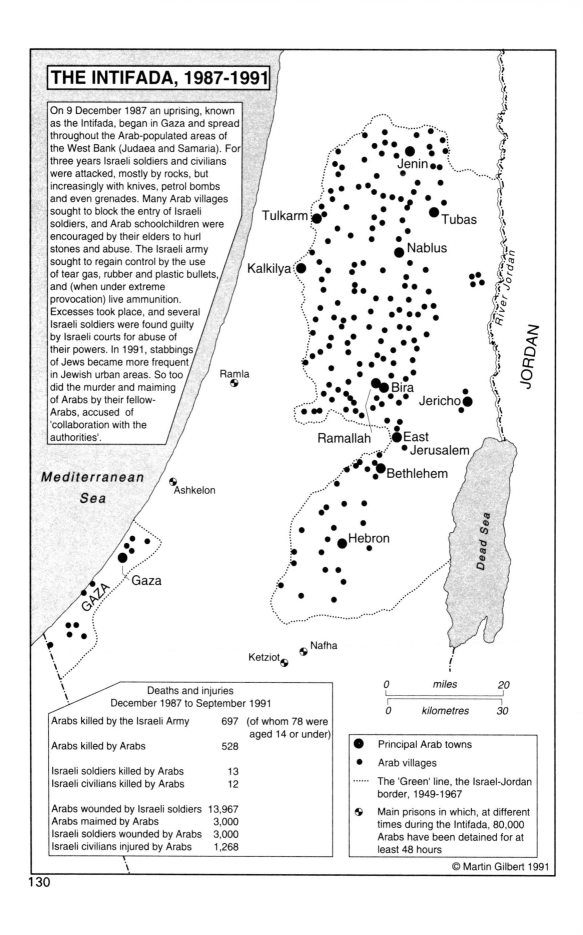

THE INTIFADA, 1987-1991

On 9 December 1987 an uprising, known as the Intifada, began in Gaza and spread throughout the Arab-populated areas of the West Bank (Judaea and Samaria). For three years Israeli soldiers and civilians were attacked, mostly by rocks, but increasingly with knives, petrol bombs and even grenades. Many Arab villages sought to block the entry of Israeli soldiers, and Arab schoolchildren were encouraged by their elders to hurl stones and abuse. The Israeli army sought to regain control by the use of tear gas, rubber and plastic bullets, and (when under extreme provocation) live ammunition. Excesses took place, and several Israeli soldiers were found guilty by Israeli courts for abuse of their powers. In 1991, stabbings of Jews became more frequent in Jewish urban areas. So too did the murder and maiming of Arabs by their fellow-Arabs, accused of 'collaboration with the authorities'.

Jenin

Tulkarm   Tubas

Nablus

Kalkilya

Ramla

Bira

Jericho

Ramallah   East Jerusalem

Bethlehem

Mediterranean Sea

Ashkelon

Hebron

Dead Sea

Gaza

GAZA

River Jordan

JORDAN

Nafha

Ketziot

Deaths and injuries December 1987 to September 1991		
Arabs killed by the Israeli Army	697	(of whom 78 were aged 14 or under)
Arabs killed by Arabs	528	
Israeli soldiers killed by Arabs	13	
Israeli civilians killed by Arabs	12	
Arabs wounded by Israeli soldiers	13,967	
Arabs maimed by Arabs	3,000	
Israeli soldiers wounded by Arabs	3,000	
Israeli civilians injured by Arabs	1,268	

0   miles   20

0   kilometres   30

●   Principal Arab towns

•   Arab villages

⋯⋯   The 'Green' line, the Israel-Jordan border, 1949-1967

◍   Main prisons in which, at different times during the Intifada, 80,000 Arabs have been detained for at least 48 hours

© Martin Gilbert 1991

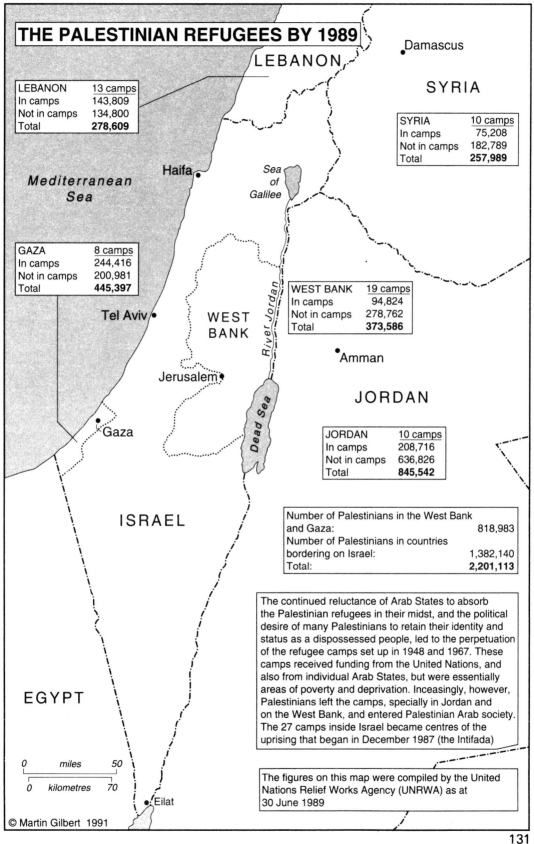

THE PALESTINIAN REFUGEES BY 1989

LEBANON

Damascus

SYRIA

LEBANON	13 camps
In camps	143,809
Not in camps	134,800
Total	**278,609**

SYRIA	10 camps
In camps	75,208
Not in camps	182,789
Total	**257,989**

Haifa

Sea of Galilee

Mediterranean Sea

GAZA	8 camps
In camps	244,416
Not in camps	200,981
Total	**445,397**

WEST BANK	19 camps
In camps	94,824
Not in camps	278,762
Total	**373,586**

River Jordan

Tel Aviv

WEST BANK

Amman

Jerusalem

Dead Sea

JORDAN

Gaza

JORDAN	10 camps
In camps	208,716
Not in camps	636,826
Total	**845,542**

Number of Palestinians in the West Bank and Gaza:	818,983
Number of Palestinians in countries bordering on Israel:	1,382,140
Total:	**2,201,113**

ISRAEL

The continued reluctance of Arab States to absorb the Palestinian refugees in their midst, and the political desire of many Palestinians to retain their identity and status as a dispossessed people, led to the perpetuation of the refugee camps set up in 1948 and 1967. These camps received funding from the United Nations, and also from individual Arab States, but were essentially areas of poverty and deprivation. Inceasingly, however, Palestinians left the camps, specially in Jordan and on the West Bank, and entered Palestinian Arab society. The 27 camps inside Israel became centres of the uprising that began in December 1987 (the Intifada)

EGYPT

The figures on this map were compiled by the United Nations Relief Works Agency (UNRWA) as at 30 June 1989

0	miles	50

0	kilometres	70

Eilat

© Martin Gilbert 1991

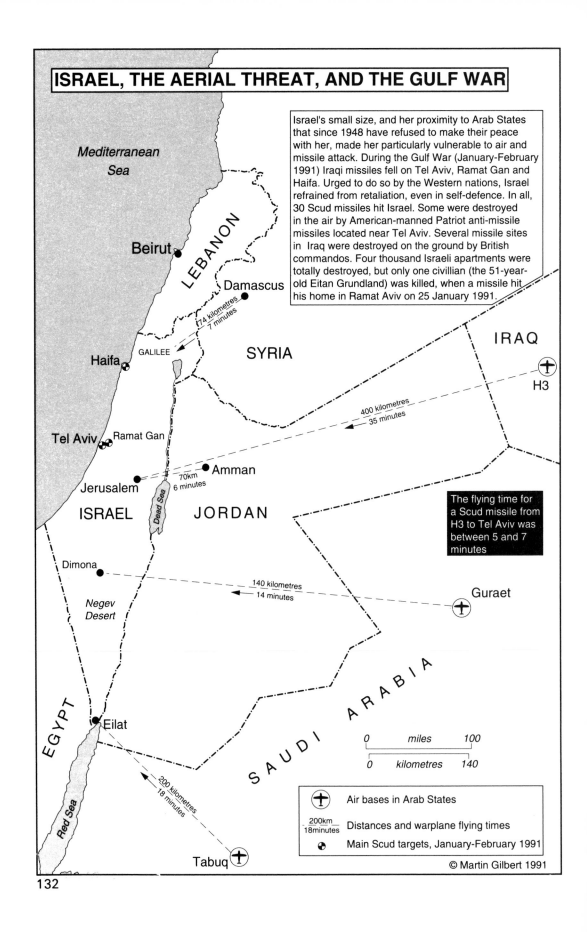

ISRAEL, THE AERIAL THREAT, AND THE GULF WAR

Mediterranean Sea

Israel's small size, and her proximity to Arab States that since 1948 have refused to make their peace with her, made her particularly vulnerable to air and missile attack. During the Gulf War (January-February 1991) Iraqi missiles fell on Tel Aviv, Ramat Gan and Haifa. Urged to do so by the Western nations, Israel refrained from retaliation, even in self-defence. In all, 30 Scud missiles hit Israel. Some were destroyed in the air by American-manned Patriot anti-missile missiles located near Tel Aviv. Several missile sites in Iraq were destroyed on the ground by British commandos. Four thousand Israeli apartments were totally destroyed, but only one civillian (the 51-year-old Eitan Grundland) was killed, when a missile hit his home in Ramat Aviv on 25 January 1991.

Beirut

LEBANON

Damascus

174 kilometres
7 minutes

GALILEE

SYRIA

IRAQ

H3

Haifa

400 kilometres
35 minutes

Tel Aviv **Ramat Gan**

Dead Sea

Amman

70km
6 minutes

Jerusalem

ISRAEL

JORDAN

The flying time for a Scud missile from H3 to Tel Aviv was between 5 and 7 minutes

Dimona

140 kilometres
14 minutes

Guraet

Negev Desert

S A U D I   A R A B I A

EGYPT

Eilat

| 0 | miles | 100 |
| 0 | kilometres | 140 |

200 kilometres
18 minutes

Red Sea

Tabuq

Air bases in Arab States

$\frac{200km}{18minutes}$ Distances and warplane flying times

Main Scud targets, January-February 1991

© Martin Gilbert 1991

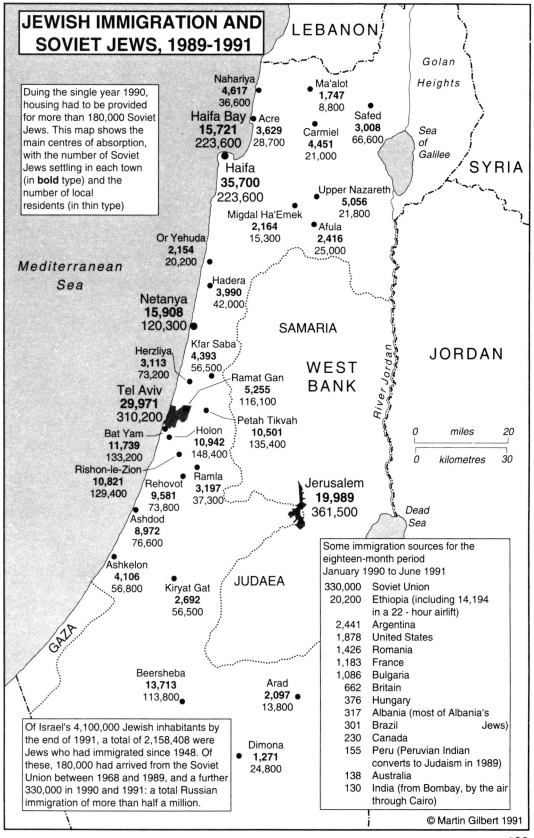

JEWISH IMMIGRATION AND SOVIET JEWS, 1989-1991

LEBANON

Golan

Heights

During the single year 1990, housing had to be provided for more than 180,000 Soviet Jews. This map shows the main centres of absorption, with the number of Soviet Jews settling in each town (in **bold** type) and the number of local residents (in thin type)

Nahariya
4,617
36,600

Ma'alot
1,747
8,800

Haifa Bay
15,721
223,600

Acre
3,629
28,700

Safed
3,008
66,600

Sea
of
Galilee

SYRIA

Carmiel
4,451
21,000

Haifa
35,700
223,600

Upper Nazareth
5,056
21,800

Migdal Ha'Emek
2,164
15,300

Afula
2,416
25,000

Or Yehuda
2,154
20,200

Mediterranean
Sea

Hadera
3,990
42,000

SAMARIA

Netanya
15,908
120,300

Kfar Saba
4,393
56,500

WEST
BANK

JORDAN

Herzliya
3,113
73,200

Ramat Gan
5,255
116,100

River Jordan

Tel Aviv
29,971
310,200

Petah Tikvah
10,501
135,400

0	miles	20

0	kilometres	30

Bat Yam
11,739
133,200

Holon
10,942
148,400

Rishon-le-Zion
10,821
129,400

Rehovot
9,581
73,800

Ramla
3,197
37,300

Jerusalem
19,989
361,500

Dead
Sea

Ashdod
8,972
76,600

Ashkelon
4,106
56,800

JUDAEA

Kiryat Gat
2,692
56,500

GAZA

Some immigration sources for the eighteen-month period
January 1990 to June 1991

330,000	Soviet Union
20,200	Ethiopia (including 14,194 in a 22 - hour airlift)
2,441	Argentina
1,878	United States
1,426	Romania
1,183	France
1,086	Bulgaria
662	Britain
376	Hungary
317	Albania (most of Albania's
301	Brazil Jews)
230	Canada
155	Peru (Peruvian Indian converts to Judaism in 1989)
138	Australia
130	India (from Bombay, by the air through Cairo)

Beersheba
13,713
113,800

Arad
2,097
13,800

Of Israel's 4,100,000 Jewish inhabitants by the end of 1991, a total of 2,158,408 were Jews who had immigrated since 1948. Of these, 180,000 had arrived from the Soviet Union between 1968 and 1989, and a further 330,000 in 1990 and 1991: a total Russian immigration of more than half a million.

Dimona
1,271
24,800

© Martin Gilbert 1991

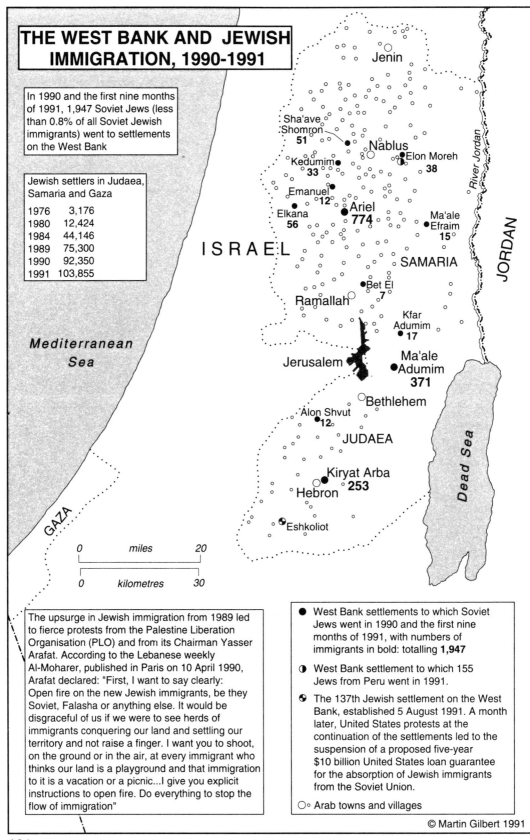

THE WEST BANK AND JEWISH IMMIGRATION, 1990-1991

In 1990 and the first nine months of 1991, 1,947 Soviet Jews (less than 0.8% of all Soviet Jewish immigrants) went to settlements on the West Bank

Jewish settlers in Judaea, Samaria and Gaza

1976	3,176
1980	12,424
1984	44,146
1989	75,300
1990	92,350
1991	103,855

Jenin

Sha'ave Shomron **51**

Kedumim **33**

Nablus

Elon Moreh **38**

Emanuel **12**

Elkana **56**

Ariel **774**

Ma'ale Efraim **15**

ISRAEL

SAMARIA

Bet El **7**

Ramallah

Kfar Adumim **17**

Jerusalem

Ma'ale Adumim **371**

Bethlehem

Alon Shvut **12**

JUDAEA

Kiryat Arba **253**

Hebron

Eshkoliot

Mediterranean Sea

Dead Sea

JORDAN

River Jordan

GAZA

| 0 | miles | 20 |
| 0 | kilometres | 30 |

The upsurge in Jewish immigration from 1989 led to fierce protests from the Palestine Liberation Organisation (PLO) and from its Chairman Yasser Arafat. According to the Lebanese weekly Al-Moharer, published in Paris on 10 April 1990, Arafat declared: "First, I want to say clearly: Open fire on the new Jewish immigrants, be they Soviet, Falasha or anything else. It would be disgraceful of us if we were to see herds of immigrants conquering our land and settling our territory and not raise a finger. I want you to shoot, on the ground or in the air, at every immigrant who thinks our land is a playground and that immigration to it is a vacation or a picnic...I give you explicit instructions to open fire. Do everything to stop the flow of immigration"

● West Bank settlements to which Soviet Jews went in 1990 and the first nine months of 1991, with numbers of immigrants in bold: totalling **1,947**

◑ West Bank settlement to which 155 Jews from Peru went in 1991.

◐ The 137th Jewish settlement on the West Bank, established 5 August 1991. A month later, United States protests at the continuation of the settlements led to the suspension of a proposed five-year $10 billion United States loan guarantee for the absorption of Jewish immigrants from the Soviet Union.

○○ Arab towns and villages

© Martin Gilbert 1991

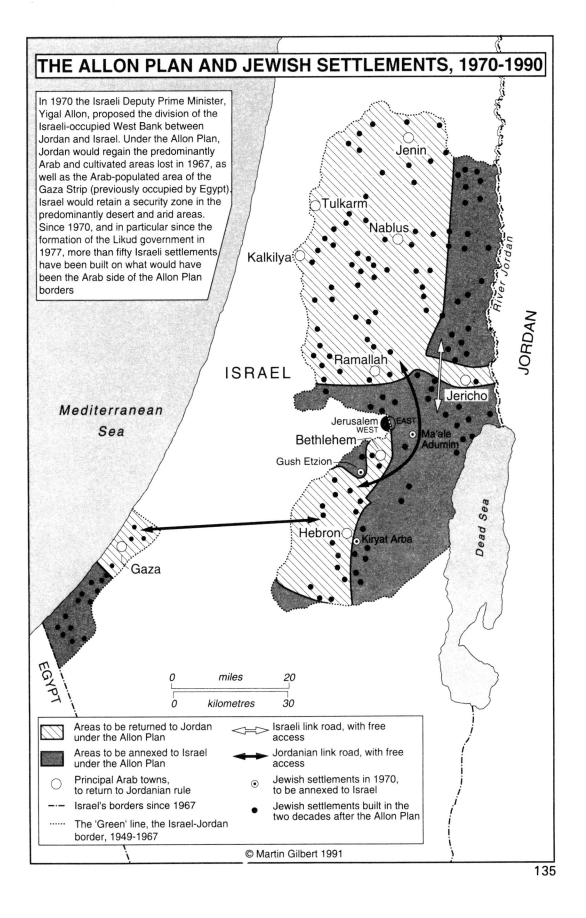

THE ALLON PLAN AND JEWISH SETTLEMENTS, 1970-1990

In 1970 the Israeli Deputy Prime Minister, Yigal Allon, proposed the division of the Israeli-occupied West Bank between Jordan and Israel. Under the Allon Plan, Jordan would regain the predominantly Arab and cultivated areas lost in 1967, as well as the Arab-populated area of the Gaza Strip (previously occupied by Egypt). Israel would retain a security zone in the predominantly desert and arid areas. Since 1970, and in particular since the formation of the Likud government in 1977, more than fifty Israeli settlements have been built on what would have been the Arab side of the Allon Plan borders

Jenin

Tulkarm

Nablus

Kalkilya

River Jordan

JORDAN

Ramallah

ISRAEL

Jericho

Mediterranean
Sea

Jerusalem EAST
WEST
Ma'ale
Adumim

Bethlehem

Gush Etzion

Dead Sea

Hebron
Kiryat Arba

Gaza

EGYPT

| 0 | miles | 20 |
| 0 | kilometres | 30 |

Areas to be returned to Jordan under the Allon Plan

Israeli link road, with free access

Areas to be annexed to Israel under the Allon Plan

Jordanian link road, with free access

○ Principal Arab towns, to return to Jordanian rule

⊙ Jewish settlements in 1970, to be annexed to Israel

–·– Israel's borders since 1967

● Jewish settlements built in the two decades after the Allon Plan

...... The 'Green' line, the Israel-Jordan border, 1949-1967

© Martin Gilbert 1991

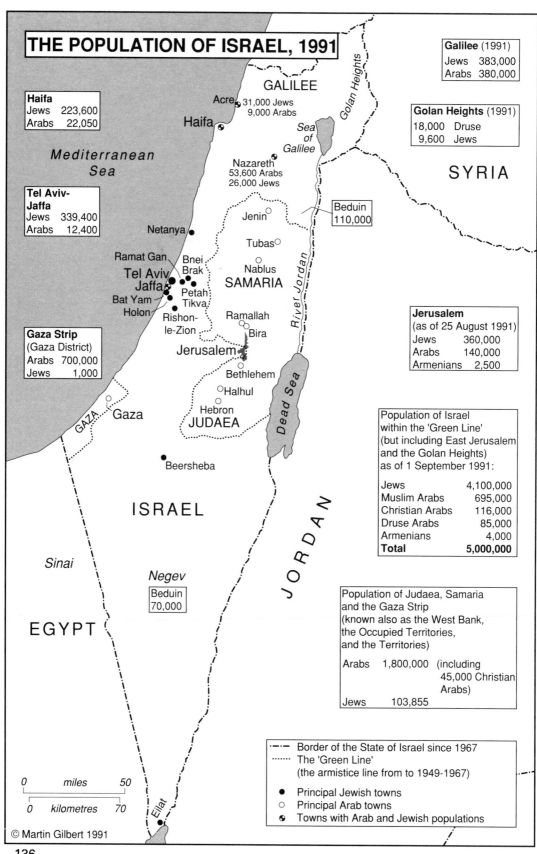

THE POPULATION OF ISRAEL, 1991

Galilee (1991)
Jews 383,000
Arabs 380,000

GALILEE

Acre ● 31,000 Jews
9,000 Arabs

Haifa
Jews 223,600
Arabs 22,050

Haifa

*Sea
of
Galilee*

Golan Heights

Golan Heights (1991)
18,000 Druse
9,600 Jews

*Mediterranean
Sea*

SYRIA

Nazareth
53,600 Arabs
26,000 Jews

**Tel Aviv-
Jaffa**
Jews 339,400
Arabs 12,400

Netanya ●

Jenin ○

Beduin
110,000

Tubas ○

Ramat Gan ● Bnei
Tel Aviv ● Brak
Jaffa ● ● Petah
Bat Yam ● Tikva
Holon ● Rishon-
le-Zion ●

Nablus ○

SAMARIA

River Jordan

Ramallah ○
Bira ○

Jerusalem
(as of 25 August 1991)
Jews 360,000
Arabs 140,000
Armenians 2,500

Jerusalem

Bethlehem ○

Gaza Strip
(Gaza District)
Arabs 700,000
Jews 1,000

○ Halhul

Hebron ○
JUDAEA

Dead Sea

GAZA ○ Gaza

Beersheba ●

Population of Israel
within the 'Green Line'
(but including East Jerusalem
and the Golan Heights)
as of 1 September 1991:

Jews	4,100,000
Muslim Arabs	695,000
Christian Arabs	116,000
Druse Arabs	85,000
Armenians	4,000
Total	**5,000,000**

ISRAEL

J O R D A N

Sinai

Negev
Beduin
70,000

EGYPT

Population of Judaea, Samaria
and the Gaza Strip
(known also as the West Bank,
the Occupied Territories,
and the Territories)

Arabs 1,800,000 (including
45,000 Christian
Arabs)
Jews 103,855

—·— Border of the State of Israel since 1967
.... The 'Green Line'
(the armistice line from to 1949-1967)

● Principal Jewish towns
○ Principal Arab towns
◐ Towns with Arab and Jewish populations

0   miles   50
0   kilometres   70

Eilat

© Martin Gilbert 1991

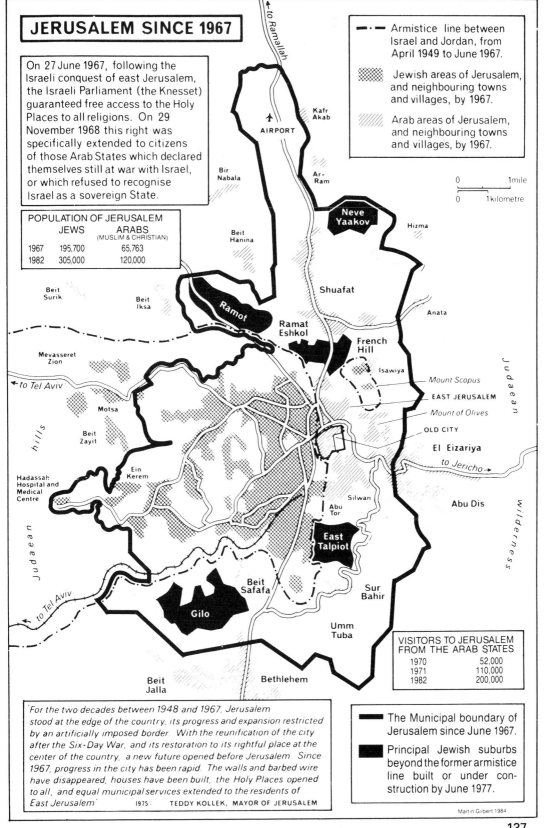

JERUSALEM SINCE 1967

On 27 June 1967, following the Israeli conquest of east Jerusalem, the Israeli Parliament (the Knesset) guaranteed free access to the Holy Places to all religions. On 29 November 1968 this right was specifically extended to citizens of those Arab States which declared themselves still at war with Israel, or which refused to recognise Israel as a sovereign State.

POPULATION OF JERUSALEM

	JEWS	ARABS (MUSLIM & CHRISTIAN)
1967	195,700	65,763
1982	305,000	120,000

- – · – Armistice line between Israel and Jordan, from April 1949 to June 1967.
- Jewish areas of Jerusalem, and neighbouring towns and villages, by 1967.
- Arab areas of Jerusalem, and neighbouring towns and villages, by 1967.

0   1 mile
0   1 kilometre

to Ramallah

Kafr Akab

AIRPORT

Bir Nabala

Ar-Ram

Beit Hanina

Neve Yaakov

Hizma

Shuafat

Anata

Beit Surik

Beit Iksa

Ramot

Ramat Eshkol

French Hill

Isawiya

Mevasseret Zion

Mount Scopus

EAST JERUSALEM

to Tel Aviv

Motsa

Mount of Olives

OLD CITY

Beit Zayit

El Eizariya

to Jericho

Ein Kerem

Hadassah Hospital and Medical Centre

Silwan

Abu Dis

Abu Tor

Judaean

wilderness

East Talpiot

Beit Safafa

Sur Bahir

to Tel Aviv

Gilo

Umm Tuba

VISITORS TO JERUSALEM FROM THE ARAB STATES

1970	52,000
1971	110,000
1982	200,000

Beit Jalla

Bethlehem

Judaean   hills

For the two decades between 1948 and 1967, Jerusalem stood at the edge of the country, its progress and expansion restricted by an artificially imposed border. With the reunification of the city after the Six-Day War, and its restoration to its rightful place at the center of the country, a new future opened before Jerusalem. Since 1967, progress in the city has been rapid. The walls and barbed wire have disappeared, houses have been built, the Holy Places opened to all, and equal municipal services extended to the residents of East Jerusalem 1975:   TEDDY KOLLEK, MAYOR OF JERUSALEM

- ▬ The Municipal boundary of Jerusalem since June 1967.
- ■ Principal Jewish suburbs beyond the former armistice line built or under construction by June 1977.

Martin Gilbert 1984

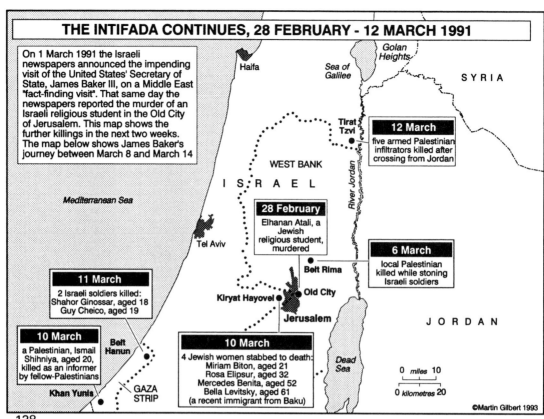

THE INTIFADA CONTINUES, 28 FEBRUARY - 12 MARCH 1991

On 1 March 1991 the Israeli newspapers announced the impending visit of the United States' Secretary of State, James Baker III, on a Middle East "fact-finding visit". That same day the newspapers reported the murder of an Israeli religious student in the Old City of Jerusalem. This map shows the further killings in the next two weeks. The map below shows James Baker's journey between March 8 and March 14

Mediterranean Sea

Haifa

Golan Heights

Sea of Galilee

S Y R I A

Tirat Tzvi

12 March
five armed Palestinian infiltrators killed after crossing from Jordan

WEST BANK

I S R A E L

River Jordan

28 February
Elhanan Atali, a Jewish religious student, murdered

Tel Aviv

6 March
local Palestinian killed while stoning Israeli soldiers

Belt Rima

11 March
2 Israeli soldiers killed: Shahor Ginossar, aged 18 Guy Cheico, aged 19

Kiryat Hayovel   Old City

Jerusalem

J O R D A N

10 March
a Palestinian, Ismail Shihniya, aged 20, killed as an informer by fellow-Palestinians

Belt Hanun

10 March
4 Jewish women stabbed to death:
Miriam Biton, aged 21
Rosa Elipsur, aged 32
Mercedes Benita, aged 52
Bella Levitsky, aged 61
(a recent immigrant from Baku)

Khan Yunis   GAZA STRIP

Dead Sea

0   miles   10
0   kilometres   20

©Martin Gilbert 1993

138

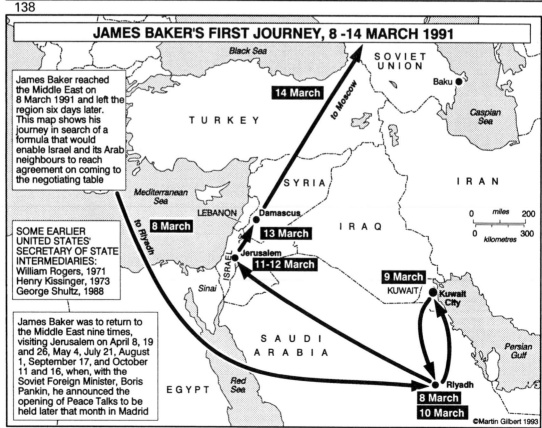

JAMES BAKER'S FIRST JOURNEY, 8 -14 MARCH 1991

Black Sea

S O V I E T U N I O N

Baku

14 March

to Moscow

James Baker reached the Middle East on 8 March 1991 and left the region six days later. This map shows his journey in search of a formula that would enable Israel and its Arab neighbours to reach agreement on coming to the negotiating table

T U R K E Y

Caspian Sea

Mediterranean Sea

S Y R I A

I R A N

to Riyadh

8 March

LEBANON   Damascus

13 March

SOME EARLIER UNITED STATES' SECRETARY OF STATE INTERMEDIARIES:
William Rogers, 1971
Henry Kissinger, 1973
George Shultz, 1988

ISRAEL   Jerusalem

11-12 March

I R A Q

0   miles   200
0   kilometres   300

Sinai

9 March
KUWAIT

Kuwait City

James Baker was to return to the Middle East nine times, visiting Jerusalem on April 8, 19 and 26, May 4, July 21, August 1, September 17, and October 11 and 16, when, with the Soviet Foreign Minister, Boris Pankin, he announced the opening of Peace Talks to be held later that month in Madrid

S A U D I A R A B I A

Persian Gulf

E G Y P T   *Red Sea*

Riyadh

8 March
10 March

©Martin Gilbert 1993

139

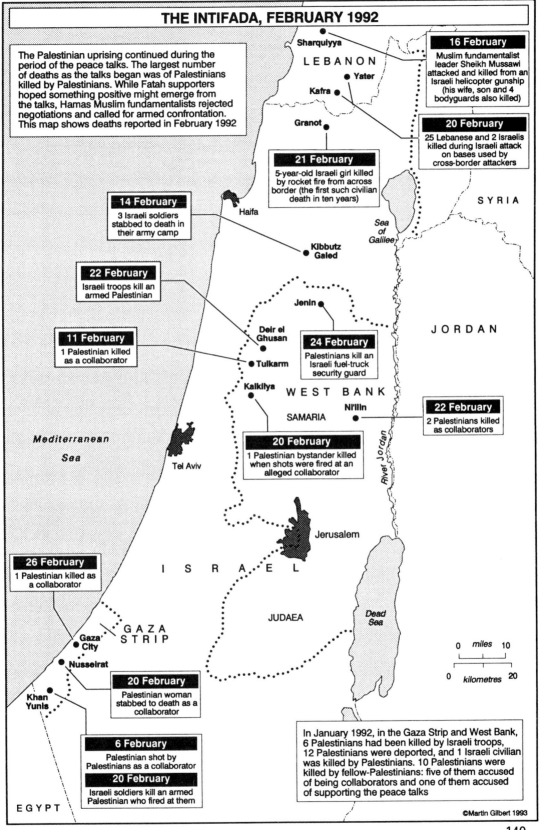

THE INTIFADA, FEBRUARY 1992

The Palestinian uprising continued during the period of the peace talks. The largest number of deaths as the talks began was of Palestinians killed by Palestinians. While Fatah supporters hoped something positive might emerge from the talks, Hamas Muslim fundamentalists rejected negotiations and called for armed confrontation. This map shows deaths reported in February 1992

16 February
Muslim fundamentalist leader Sheikh Mussawi attacked and killed from an Israeli helicopter gunship (his wife, son and 4 bodyguards also killed)

20 February
25 Lebanese and 2 Israelis killed during Israeli attack on bases used by cross-border attackers

21 February
5-year-old Israeli girl killed by rocket fire from across border (the first such civilian death in ten years)

14 February
3 Israeli soldiers stabbed to death in their army camp

22 February
Israeli troops kill an armed Palestinian

11 February
1 Palestinian killed as a collaborator

24 February
Palestinians kill an Israeli fuel-truck security guard

22 February
2 Palestinians killed as collaborators

20 February
1 Palestinian bystander killed when shots were fired at an alleged collaborator

26 February
1 Palestinian killed as a collaborator

20 February
Palestinian woman stabbed to death as a collaborator

6 February
Palestinian shot by Palestinians as a collaborator

20 February
Israeli soldiers kill an armed Palestinian who fired at them

In January 1992, in the Gaza Strip and West Bank, 6 Palestinians had been killed by Israeli troops, 12 Palestinians were deported, and 1 Israeli civilian was killed by Palestinians. 10 Palestinians were killed by fellow-Palestinians: five of them accused of being collaborators and one of them accused of supporting the peace talks

LEBANON
Sharqulyya
Yater
Kafra
Granot
SYRIA
Haifa
Sea of Galilee
Kibbutz Galed
Jenin
JORDAN
Deir el Ghusan
Tulkarm
Kalkilya
WEST BANK
Ni'llin
SAMARIA
Mediterranean Sea
Tel Aviv
River Jordan
Jerusalem
ISRAEL
JUDAEA
Dead Sea
GAZA STRIP
Gaza City
Nusseirat
Khan Yunis
EGYPT

0 miles 10
0 kilometres 20

©Martin Gilbert 1993

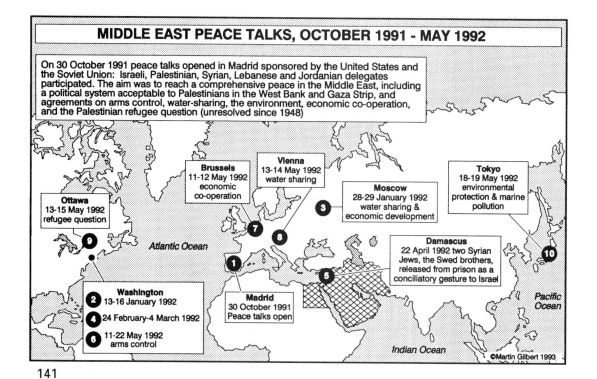

MIDDLE EAST PEACE TALKS, OCTOBER 1991 - MAY 1992

On 30 October 1991 peace talks opened in Madrid sponsored by the United States and the Soviet Union: Israeli, Palestinian, Syrian, Lebanese and Jordanian delegates participated. The aim was to reach a comprehensive peace in the Middle East, including a political system acceptable to Palestinians in the West Bank and Gaza Strip, and agreements on arms control, water-sharing, the environment, economic co-operation, and the Palestinian refugee question (unresolved since 1948)

Brussels
11-12 May 1992
economic
co-operation

Vienna
13-14 May 1992
water sharing

Tokyo
18-19 May 1992
environmental
protection & marine
pollution

Ottawa
13-15 May 1992
refugee question

Moscow
28-29 January 1992
water sharing &
economic development

Atlantic Ocean

Damascus
22 April 1992 two Syrian
Jews, the Swed brothers,
released from prison as a
conciliatory gesture to Israel

Pacific Ocean

Washington
2 13-16 January 1992
4 24 February-4 March 1992
6 11-22 May 1992
    arms control

Madrid
30 October 1991
Peace talks open

Indian Ocean

©Martin Gilbert 1993

141

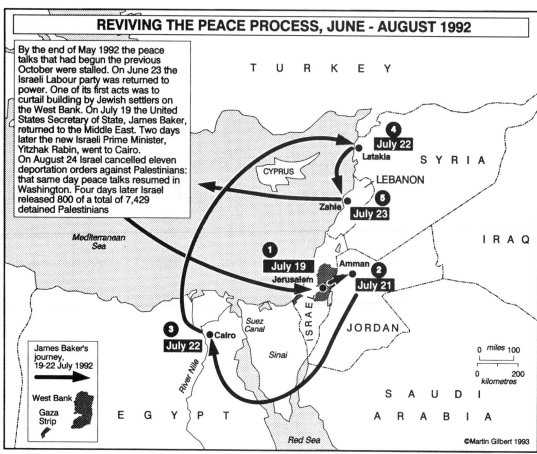

REVIVING THE PEACE PROCESS, JUNE - AUGUST 1992

By the end of May 1992 the peace talks that had begun the previous October were stalled. On June 23 the Israeli Labour party was returned to power. One of its first acts was to curtail building by Jewish settlers on the West Bank. On July 19 the United States Secretary of State, James Baker, returned to the Middle East. Two days later the new Israeli Prime Minister, Yitzhak Rabin, went to Cairo. On August 24 Israel cancelled eleven deportation orders against Palestinians: that same day peace talks resumed in Washington. Four days later Israel released 800 of a total of 7,429 detained Palestinians

T U R K E Y

4 **July 22**
Latakia

S Y R I A

CYPRUS

LEBANON

5 **July 23**
Zahle

I R A Q

Mediterranean Sea

1 **July 19**
Jerusalem

Amman
2 **July 21**

J O R D A N

3 **July 22**
Cairo

Suez Canal

Sinai

0 miles 100
0 kilometres 200

James Baker's journey,
19-22 July 1992

West Bank

Gaza Strip

E G Y P T

River Nile

S A U D I

A R A B I A

Red Sea

©Martin Gilbert 1993

142

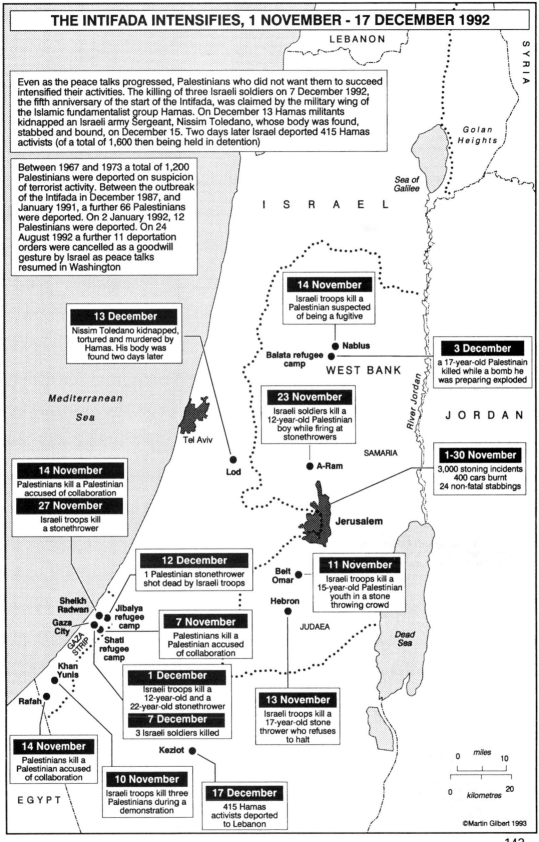

THE INTIFADA INTENSIFIES, 1 NOVEMBER - 17 DECEMBER 1992

LEBANON

SYRIA

Even as the peace talks progressed, Palestinians who did not want them to succeed intensified their activities. The killing of three Israeli soldiers on 7 December 1992, the fifth anniversary of the start of the Intifada, was claimed by the military wing of the Islamic fundamentalist group Hamas. On December 13 Hamas militants kidnapped an Israeli army Sergeant, Nissim Toledano, whose body was found, stabbed and bound, on December 15. Two days later Israel deported 415 Hamas activists (of a total of 1,600 then being held in detention)

Golan Heights

Between 1967 and 1973 a total of 1,200 Palestinians were deported on suspicion of terrorist activity. Between the outbreak of the Intifada in December 1987, and January 1991, a further 66 Palestinians were deported. On 2 January 1992, 12 Palestinians were deported. On 24 August 1992 a further 11 deportation orders were cancelled as a goodwill gesture by Israel as peace talks resumed in Washington

Sea of Galilee

I S R A E L

14 November
Israeli troops kill a Palestinian suspected of being a fugitive

● Nablus

13 December
Nissim Toledano kidnapped, tortured and murdered by Hamas. His body was found two days later

Balata refugee camp ●

WEST BANK

3 December
a 17-year-old Palestinain killed while a bomb he was preparing exploded

Mediterranean Sea

River Jordan

J O R D A N

23 November
Israeli soldiers kill a 12-year-old Palestinian boy while firing at stonethrowers

Tel Aviv

SAMARIA

1-30 November
3,000 stoning incidents
400 cars burnt
24 non-fatal stabbings

● A-Ram

14 November
Palestinians kill a Palestinian accused of collaboration

Lod ●

27 November
Israeli troops kill a stonethrower

Jerusalem

12 December
1 Palestinian stonethrower shot dead by Israeli troops

Belt Omar ●

11 November
Israeli troops kill a 15-year-old Palestinian youth in a stone throwing crowd

Hebron ●

Sheikh Radwan ●
Jibalya refugee camp ●
Gaza City ●
Shati refugee camp ●

7 November
Palestinians kill a Palestinian accused of collaboration

JUDAEA

Dead Sea

Khan Yunis ●

1 December
Israeli troops kill a 12-year-old and a 22-year-old stonethrower

Rafah ●

7 December
3 Israeli soldiers killed

13 November
Israeli troops kill a 17-year-old stone thrower who refuses to halt

14 November
Palestinians kill a Palestinian accused of collaboration

Kezlot ●

10 November
Israeli troops kill three Palestinians during a demonstration

17 December
415 Hamas activists deported to Lebanon

E G Y P T

GAZA STRIP

miles
0 — 10

0 — 20
kilometres

©Martin Gilbert 1993

143

THE DECEMBER 1992 DEPORTATION AND ITS AFTERMATH

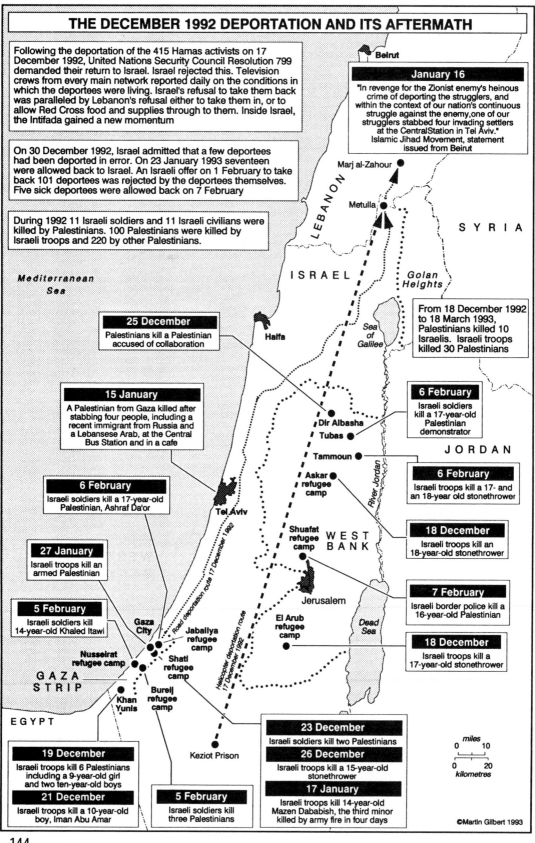

Following the deportation of the 415 Hamas activists on 17 December 1992, United Nations Security Council Resolution 799 demanded their return to Israel. Israel rejected this. Television crews from every main network reported daily on the conditions in which the deportees were living. Israel's refusal to take them back was paralleled by Lebanon's refusal either to take them in, or to allow Red Cross food and supplies through to them. Inside Israel, the Intifada gained a new momentum

On 30 December 1992, Israel admitted that a few deportees had been deported in error. On 23 January 1993 seventeen were allowed back to Israel. An Israeli offer on 1 February to take back 101 deportees was rejected by the deportees themselves. Five sick deportees were allowed back on 7 February

During 1992 11 Israeli soldiers and 11 Israeli civilians were killed by Palestinians. 100 Palestinians were killed by Israeli troops and 220 by other Palestinians.

Beirut

January 16
"In revenge for the Zionist enemy's heinous crime of deporting the strugglers, and within the context of our nation's continuous struggle against the enemy, one of our strugglers stabbed four invading settlers at the Central Station in Tel Aviv."
Islamic Jihad Movement, statement issued from Beirut

Marj al-Zahour

Metulla

S Y R I A

L E B A N O N

I S R A E L

Golan Heights

Mediterranean Sea

Sea of Galilee

From 18 December 1992 to 18 March 1993, Palestinians killed 10 Israelis. Israeli troops killed 30 Palestinians

25 December
Palestinians kill a Palestinian accused of collaboration

Haifa

15 January
A Palestinian from Gaza killed after stabbing four people, including a recent immigrant from Russia and a Lebansese Arab, at the Central Bus Station and in a cafe

Dir Albasha
Tubas

6 February
Israeli soldiers kill a 17-year-old Palestinian demonstrator

Tammoun

Askar refugee camp

J O R D A N

6 February
Israeli soldiers kill a 17-year-old Palestinian, Ashraf Da'or

Tel Aviv

River Jordan

6 February
Israeli troops kill a 17- and an 18-year old stonethrower

27 January
Israeli troops kill an armed Palestinian

Shuafat refugee camp

W E S T B A N K

18 December
Israeli troops kill an 18-year-old stonethrower

5 February
Israeli soldiers kill 14-year-old Khaled Itawi

Jerusalem

7 February
Israeli border police kill a 16-year-old Palestinian

Gaza City

Jaballya refugee camp

El Arub refugee camp

Dead Sea

18 December
Israeli troops kill a 17-year-old stonethrower

Nusseirat refugee camp

Shati refugee camp

G A Z A S T R I P

Burelj refugee camp

Khan Yunis

E G Y P T

Road deportation route 17 December 1992

Helicopter deportation route 17 December 1992

Keziot Prison

19 December
Israeli troops kill 6 Palestinians including a 9-year-old girl and two ten-year-old boys

21 December
Israeli troops kill a 10-year-old boy, Iman Abu Amar

5 February
Israeli soldiers kill three Palestinians

23 December
Israeli soldiers kill two Palestinians

26 December
Israeli troops kill a 15-year-old stonethrower

17 January
Israeli troops kill 14-year-old Mazen Dababish, the third minor killed by army fire in four days

miles
0   10

0   20
kilometres

©Martin Gilbert 1993

144

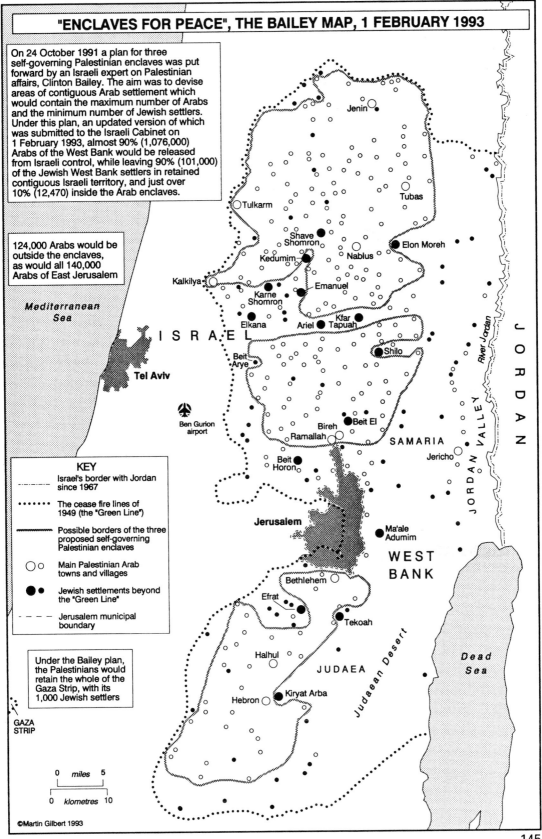

"ENCLAVES FOR PEACE", THE BAILEY MAP, 1 FEBRUARY 1993

On 24 October 1991 a plan for three
self-governing Palestinian enclaves was put
forward by an Israeli expert on Palestinian
affairs, Clinton Bailey. The aim was to devise
areas of contiguous Arab settlement which
would contain the maximum number of Arabs
and the minimum number of Jewish settlers.
Under this plan, an updated version of which
was submitted to the Israeli Cabinet on
1 February 1993, almost 90% (1,076,000)
Arabs of the West Bank would be released
from Israeli control, while leaving 90% (101,000)
of the Jewish West Bank settlers in retained
contiguous Israeli territory, and just over
10% (12,470) inside the Arab enclaves.

124,000 Arabs would be
outside the enclaves,
as would all 140,000
Arabs of East Jerusalem

*Mediterranean
Sea*

I S R A E L

Tel Aviv

Jenin

Tulkarm

Shave
Shomron

Kedumim

Nablus

Elon Moreh

Kalkilya

Karne
Shomron

Emanuel

Elkana

Ariel

Kfar
Tapuah

Tubas

Shilo

Beit
Arye

Ben Gurion
airport

Beit El

Bireh

Ramallah

River Jordan

J O R D A N

J O R D A N   V A L L E Y

S A M A R I A

Jericho

Beit
Horon

Jerusalem

Ma'ale
Adumim

W E S T
B A N K

Bethlehem

Efrat

Tekoah

Judaean Desert

*Dead
Sea*

Halhul

J U D A E A

Hebron

Kiryat Arba

KEY

— · — · — Israel's border with Jordan
since 1967

· · · · · · · The cease fire lines of
1949 (the "Green Line")

─────── Possible borders of the three
proposed self-governing
Palestinian enclaves

○ ◦ Main Palestinian Arab
towns and villages

● • Jewish settlements beyond
the "Green Line"

— — — Jerusalem municipal
boundary

Under the Bailey plan,
the Palestinians would
retain the whole of the
Gaza Strip, with its
1,000 Jewish settlers

GAZA
STRIP

0   miles   5

0   kilometres   10

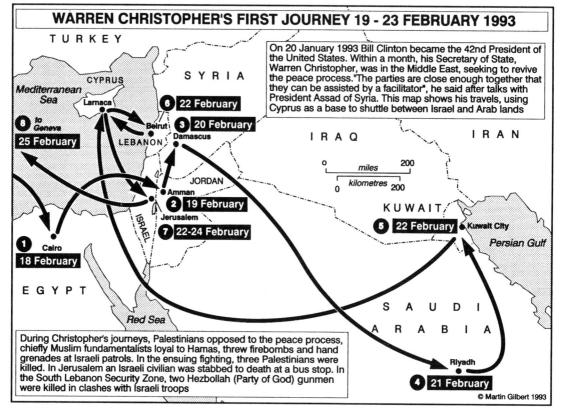

WARREN CHRISTOPHER'S FIRST JOURNEY 19 - 23 FEBRUARY 1993

TURKEY

CYPRUS

SYRIA

Mediterranean Sea

Larnaca

6 22 February

Beirut

3 20 February

LEBANON

Damascus

8 to Geneva

25 February

IRAQ

IRAN

On 20 January 1993 Bill Clinton became the 42nd President of the United States. Within a month, his Secretary of State, Warren Christopher, was in the Middle East, seeking to revive the peace process."The parties are close enough together that they can be assisted by a facilitator", he said after talks with President Assad of Syria. This map shows his travels, using Cyprus as a base to shuttle between Israel and Arab lands

JORDAN

Amman

2 19 February

ISRAEL

Jerusalem

7 22-24 February

1 Cairo

18 February

EGYPT

Red Sea

0   miles   200

0   kilometres   200

KUWAIT

5 22 February

Kuwait City

Persian Gulf

SAUDI

ARABIA

Riyadh

4 21 February

During Christopher's journeys, Palestinians opposed to the peace process, chiefly Muslim fundamentalists loyal to Hamas, threw firebombs and hand grenades at Israeli patrols. In the ensuing fighting, three Palestinians were killed. In Jerusalem an Israeli civilian was stabbed to death at a bus stop. In the South Lebanon Security Zone, two Hezbollah (Party of God) gunmen were killed in clashes with Israeli troops

© Martin Gilbert 1993

WITHDRAWN